P9-BAU-782

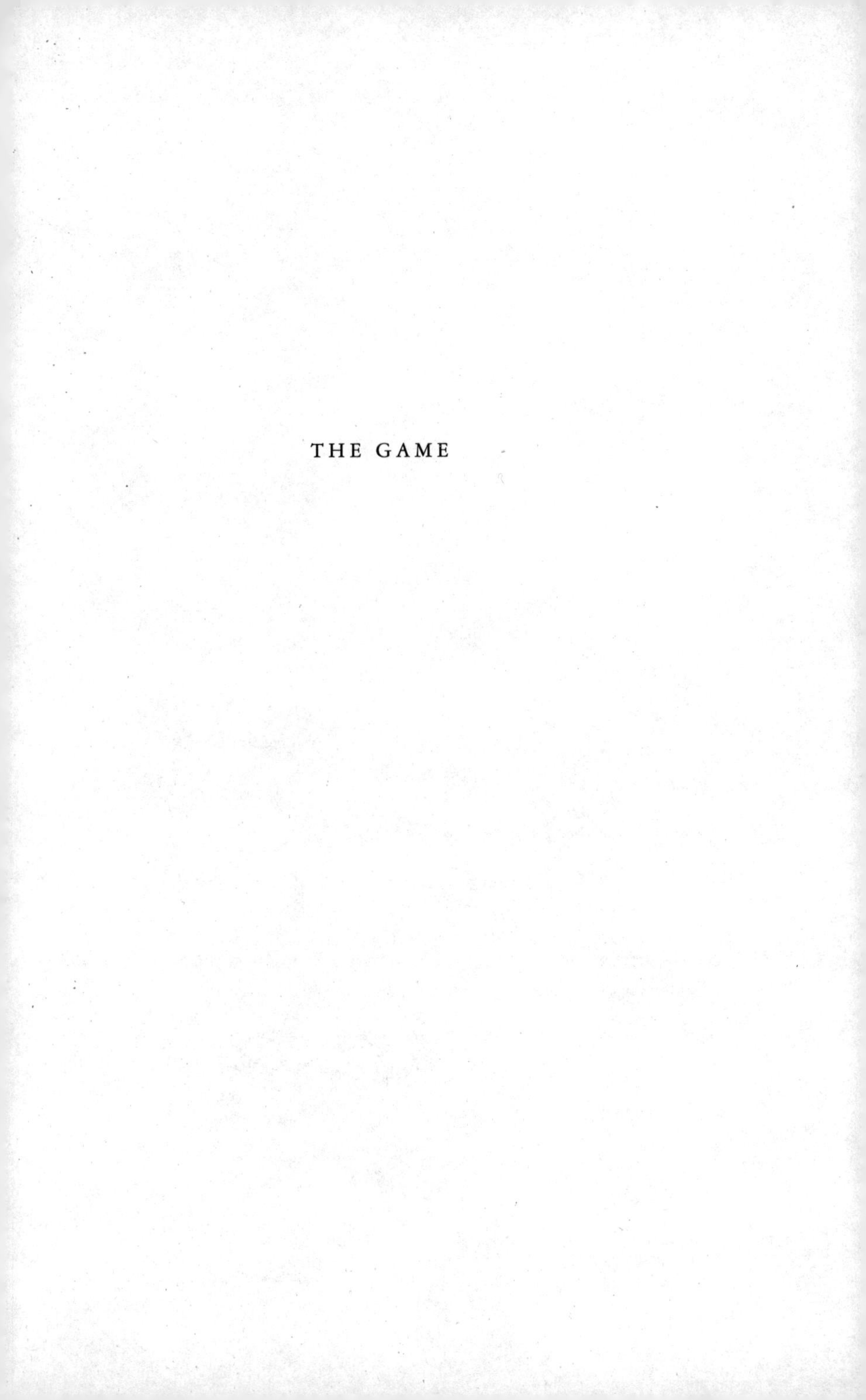

THE GAME

McSWEENEY'S
SAN FRANCISCO

Map design by Luigi Farrauto and Andrea Novali of 100km Studio.

ISBN 978-1-952119-00-2

10 9 8 7 6 5 4 3 2 1

www.mcsweeneys.net

Printed in Canada

THE GAME

A Digital Turning Point

ALESSANDRO BARICCO

TRANSLATED BY CLARISSA BOTSFORD

McSWEENEY'S
SAN FRANCISCO

To Carlo, Oscar, and Andrea.
To the seven wise men.
To those who invent the Holden School every day.
This lesson is for you.

The Game

▸ Username
Password

Play

Maps

Level Up

USERNAME

Ten years or so ago, I wrote a book called *The Barbarians*. At the time, many people—from a whole range of cultural backgrounds—were beginning to comment negatively on the disconcerting fact that some of the most noble, significant, and striking advances mankind had made over previous centuries were increasingly devalued, neglected, or oversimplified. Human beings appeared to have forgotten how to do things properly, whether it was eating, studying, traveling, having fun, or making love. That is, with the knowledge, care, and attention to detail that they had acquired through generations. It looked as if they preferred to do things quickly and superficially.

Watching kids go about their daily activities upset parents in particular. They felt their offspring had fallen prey to an inexplicable genetic reversal which meant that, rather than advancing the species, they were quite clearly triggering a mysterious regression. Always attached to a device of some kind, incapable of focusing, disoriented by sterile multitasking, parents felt their kids were skimming the surface of things without any discernible aim, except perhaps to reduce their chances of suffering. In their view, their offspring's incomprehensible way of going about things heralded an imminent crisis: a cultural apocalypse looming on the horizon.

It was all very irritating. People's intelligence seemed to have been momentarily reduced to denouncing the demise of this or that feature of old. They spent all their time defending things that, in their view, were degenerating. People who were usually quite sensible suddenly found themselves signing petitions in favor of old-fashioned dairies or the English subjunctive without any sense of the ridiculousness of their actions. Defending something, making sure it was not swept away by the winds of change, gave them a feeling of superiority. Most people legitimately felt no obligation toward the future; saving the past was more urgent.

I must add that a ready-made explanation for civilization caving in in this way was at hand. The picture wasn't entirely clear, but people were pretty sure that two factors in particular were chiefly responsible: the digital revolution (all those computers) and globalization (all those buyers and sellers). In the incubators of these two irresistible forces, they claimed, a type of human being had been hatched whose ambitions they didn't understand, whose language they didn't speak, whose tastes they didn't appreciate, and whose manners they deplored. They called these new human beings "barbarians." It was a term that had already been used in our history as dominators of the planet; a useful short-cut to mean annoyingly different people whom we are unable to control or comprehend.

Their instinct was to keep them back. The widely held bias was that barbarians destroyed things. Period.

Hm, I thought.

In fact, I went on to write that book, and I did so to clarify to myself and to others that what we were experiencing was not an invasion of barbarians who were wiping out our refined civilization. Rather, I argued, it was a mutation that affected all of us and that would soon generate a new civilization that was in some ways better than the one we had grown up in. I was convinced it was not a destructive invasion. It was a clever mutation: a collective conversion to new survival techniques; a brilliant, strategic change of tack. My mind went back to all those amazing changes of direction to which we gave names such as Humanism, Enlightenment, and Romanticism, and I realized we were going through a similarly formidable paradigm shift. We were giving our principles a 180-degree turn, just as we had done in other historical circumstances that would later turn out to be memorable. There was no need to be scared of change; everything would work out fine. As surprising as it may have seemed, I felt we would quickly find a good reason for no longer getting our milk from the dairy and even, if need be, stop using the English subjunctive.

It wasn't foolish optimism, as I've often tried to explain. It was pure and simple realism. When people believe they are witnessing cultural decline when a sixteen-year-old no longer uses the English subjunctive without taking into account the fact that the same kid has seen thirty times the number of movies his father had seen at his age, it's not me being an optimist, it's them missing a point. When intellectuals point their critical radar at the mediocrity of the best-selling book of the moment, claiming it heralds our cultural annihilation, I try to stick to the facts. I point out that sixty years ago, the readers who would have put that book at the top of the best-seller list not only never read books, they were *illiterate*. There has plainly been some progress. In a landscape of this kind, it is never easy to establish clearly who is telling stories. Is it me, with my pedantic realism, or is it them, with their poetic penchant for doomsday fantasies? While we were wasting time debating these things, other human beings—mostly in California; mostly belonging to a relatively inconspicuous, pragmatic elite; endowed with a strong business sense—were going ahead and changing the world. They were changing the world TECHNOLOGICALLY, without bothering to explain what project they had in mind for humanity, and perhaps without considering the consequences the changes would have on our brains and sentiments. They had no opinion whatsoever about dairies or the English subjunctive. They legitimately felt that saving the past was not their task; the urgency was inventing the future.

> I came to realize, unjustifiably late in the day, that the paradigm of civilization decline is, for many people, a comforting one. It is their favorite playing field. I don't mean tragedies or catastrophes—these are the preferred habitat of small groups of unusually smart people. I'm talking about something more subtle. Though it may seem absurd, we humans are just like those animals that lay their eggs where they can safely count on some kind of ELEGANTLY SLOW DECLINE. The most common type of human intelligence—long-suffering, stalwart, patient rather than imaginative, basically conservative—is happier with the idea of unexceptional misfortune, a slightly tilted downward slope. Since it is easier to perceive the world when it passes by at a moderate pace, these people try to slow it down further; since they prefer playing defense, they are at their best in the presence of enemies and looming tragedies; since they are not programmed to attack, they fear the future.
>
> Thus, humans tend to avoid prolonged exposure to the open field of invention if they can. When they have the chance, they generally encourage their tribe to play a game better suited to their skill set: safeguarding memory. Under the shelter of the things we have safeguarded, we can rest, deposit our eggs, and wait for the future to hatch, postponing our hunger, and therefore the need to leave our den and search for food, for as long as possible.

In any case, I finally set my mind to writing this book, and I ended up publishing it in installments in a newspaper, which seemed a splendidly barbarian way of doing things. I thought of calling it *The Mutation*. However, the editor—a genius of the category—sat for a long time looking at the title and then quite simply said, "No. *The Barbarians* is much better." I'm generally good natured, so I called the book *The Barbarians*, adding a subtitle: *An Essay on Mutation*.

And that was it.

The first thing that happened took me by surprise. It was incredibly hard to convince people that my book was not AGAINST barbarians. Everyone was so keen to hear a brilliant, convincing argument in favor of the thesis that our civilization was caving in, and that THOSE PEOPLE OVER THERE were the culprits, that, as soon as they saw the title, their minds shifted to reading that our civilization was caving in and that THOSE PEOPLE OVER THERE were the culprits, no matter what words were actually on the page.

I swear.

However often I repeated—as I had explained in the book—that there were no barbarians, that we were the ones who were effecting spectacular change, people would come up to me and thank me for denouncing the damage THOSE PEOPLE OVER THERE were perpetrating. I probably should have chosen a title like *Up with the Barbarians*, though I'm still not certain it would have been enough. If a creature is protecting her eggs in her den, surrounded by all the things she has safeguarded under the warm protection of an elegant decline, it's not easy to shift her. Collective inertia decided to accept all the smug exposés: an apocalypse of some kind or other was coming to wipe out all the good things in life. Shifting ideas like these was tremendously difficult, almost impossible.

Ten years have gone by, and I can now cite one shift that has taken place in the meantime which reassures me: the collective narration has changed. The tribes have come out of their dens, and there are few people left who still think they can explain what is happening with a legend of barbarians attacking our fortresses, spurred on by ransacking merchants seeking war booty. Most humans in developed societies now accept the fact that they are living through a kind of revolution—definitely a technical revolution but perhaps also a mental one—which is destined to alter nearly all their daily actions, maybe their priorities, and almost certainly the very concept of what experience should be. They may fear the consequences, they may understand very little, but the majority now believes the revolution is both necessary and irreversible. They are also convinced that it was undertaken with the intention of addressing some of the errors of the past that cost us so dearly. It is now perceived as a task or a challenge. There are even some who think it will lead to a better world. Of course, there are still many people sitting comfortably in the shade of the declining civilization paradigm. However, little by little, like sands in an hourglass, they filter through the bottleneck of their fear and join the converted on the other side.

What happened, you may well ask, to change our minds in such a short period of time? Why have we started to accept the idea of a revolution that bets against everything we have ever known?

I don't have a precise answer. What I do have is a short list of things that twenty years ago didn't exist and now do.

- ☐ Wikipedia
- ☐ Facebook
- ☐ Skype
- ☐ YouTube
- ☐ Spotify
- ☐ Netflix

- ☐ Twitter
- ☐ YouPorn
- ☐ Airbnb
- ☐ iPhone
- ☐ Instagram
- ☐ Uber
- ☐ WhatsApp
- ☐ Tinder
- ☐ TripAdvisor
- ☐ Pinterest

If you have nothing better to do, put an X by the things to which you dedicate a not entirely insignificant portion of your day, every day.

Quite a few of them, right? It makes one wonder how on earth we filled our days before.

Doing puzzles of the Swiss Alps, perchance?

The list teaches us a great deal, but only one thing in particular is relevant here. That is that, over the past twenty years, the revolution has become embedded in our normality, in our everyday life, in the simplest of actions, in the way we deal with our desires and fears. With this level of penetration, of course, it would be stupid to deny its existence. On the other hand, to present the revolution as a metamorphosis imposed from on high by the forces of evil is also beginning to be problematic. Essentially, we have all realized that, in the most basic habits of our daily life, we have assimilated the physical and mental actions that only twenty years ago we found hard to accept in the younger generation and denounced as degenerate. What has happened? Have we been taken over? Has someone imposed an alien way of life on us?

It would be unfair to answer in the affirmative. Someone may have PROPOSED it, but every day we go back and accept the proposal, and we have made a clear U-turn in the way we live our lives. Twenty years ago, the mental posture we adopt today would have seemed grotesque, deformed, and barbarian; nowadays, it is our way of being comfortable, alive, even *elegant* in the flow of our everyday lives. The idea that we've been invaded has been dispelled. The prevailing sensation is that we have leaned out beyond the world, as we once knew it, and are beginning to colonize areas of ourselves that we have never explored before or have not yet generated. An idea of AUGMENTED HUMANITY has started to take hold. The prospect of taking part in it has turned out to be more riveting than the fear of being deported there used to be. The result is that we have submitted ourselves to a mutation whose existence we used to openly deny. We have used our intelligence to exploit it rather than boycott it. I would add that this has led us to see the end of dairies as inevitable collateral damage. In a very short space of time, we have started to open cafés that *reference vintage dairies*. It's our way of seeing the past off by metabolizing it.

We have thus clarified the situation and corrected a few basic mistakes. We now acknowledge that it is a revolution, and we are willing to believe that it is the fruit of collective action—even of collective REVINDICATION—rather than an unexpected systemic degeneration or the diabolical plan of some evil genius. We are living a future that we have extorted from the past, that is our right, and that we strongly desired. This new world is ours—this revolution is ours.

OK.

Now we need to focus on a point that is really quite interesting: IT'S A WORLD WE HAVE NO IDEA HOW TO EXPLAIN. IT'S A

REVOLUTION WHOSE ORIGIN OR PURPOSE WE WILL NEVER KNOW FOR SURE.

Well, I suppose some people may have an idea or two. However, in general, we know very little about the mutation we are triggering. Our actions have already changed at breakneck speed, but our thoughts seem to be lagging. We are behind in the task of naming everything we create in real time. Already for a while now, space and time have been out of sync. The same is true for the mental loci that for many years we referred to as "past," "soul," "individual," "experience," and "freedom." The meaning of today's concepts, such as "everything" and "nothing," would have seemed inexact only five years ago. What for centuries we called "works of art" are now without a name. What we do know for sure is that we will be orienting ourselves with maps that have not yet been drawn; we will have an idea of beauty that we cannot yet predict; we will call a web of characters that in the past we would have considered figments of our imagination "the truth." We tell ourselves that everything that is happening must have an origin and a purpose, but we have no idea what these might be. In a few centuries, we will be remembered as the conquistadores of a land where, already today, we hardly remember our way home.

Isn't it amazing?

I think it is, and that's why I'm writing this book. I'm attracted to the idea of going to live for a while in the place where the revolution we are carrying out fades, goes quiet, and sinks below the surface. Where we cannot understand its actions, where it conceals the meaning of its movements, where it denies access to the roots it is creating. Where it looks like a mysterious frontier. Where there are endless plains, with not a single smoking chimney on the horizon. No road signs. A pioneer's story.

I don't want to give the wrong impression that I have all the answers and that I'm here to explain.

I do have some maps, though. Of course, until I start my journey, I have no way of knowing whether they are reliable, accurate, or useful.

I'm writing this book in order to undertake the journey.

I don't want to waste time, so I've decided to use a compass that has never let me down: fear. Tracking the footsteps of fear—your own and other people's—always leads you to your destination. In this instance, it's easy because there are so many different fears around, and some of them are by no means unfounded.

Let's look at an example. One of these fears goes more or less like this: WE ARE FORGING AHEAD WITH OUR HEADLIGHTS OFF. It's quite true. We don't really know what triggered the revolution, and we know even less about what it aims to achieve. We have no idea of its goals, and we wouldn't be able to say with sufficient precision what its values and principles are. We know what the Enlightenment stood for, but we don't know—at least, not clearly enough—what this revolution represents. Thus, if a young person asks an older person where we are heading, the response tends to be evasive ("You tell me" is currently the best answer; you can see why it's so urgent that someone writes this book, even someone who's not me).

ANOTHER FEAR SOUNDS A BIT LIKE THIS: are we sure that this isn't the kind of technological revolution that blindly imposes an anthropological metamorphosis that cannot be controlled? We have chosen our tools, and we like using them, but did anyone—before the fact—bother to calculate the consequences on the ways we lead our lives, perhaps on our intelligence, and, in the extreme, on our very idea of good and evil? Do Gates, Bezos, Zuckerberg, Brin, Page *et al.* have a project for humanity in mind, or have they simply produced brilliant business ideas that have involuntarily, and rather randomly, produced a new kind of human being?

A FEAR THAT I LIKE IN PARTICULAR IS THIS: we are creating a truly brilliant , fairly pleasant civilization but it appears unequipped for dealing with reality. Our civilization is frivolous, but the world around us and history are not. Once we have dismantled our capacity for patience and hard, painstaking work, won't future generations be incapable of dealing with the twists of destiny? Will they be able to survive the inevitable violence of any destiny? We are beginning to think that everyone is so intent on honing their soft skills that they are losing the muscular strength required for the physical struggle against reality. Hence, the tendency to skirt around, or avoid, reality altogether and to replace it with superficial representations that adapt their content to our devices and to the type of intelligence that has developed from their logic. Are we sure this isn't a suicidal strategy? THERE IS ANOTHER STILL SUBTLER FEAR that is widespread and that I can only summarize with the following very simple words. Every day that goes by, human beings are shedding some aspect of their humanity, seemingly preferring any form of artificiality with a higher threshold of performativity and a lower limit of fallibility. Whenever they can, they delegate choices, decisions, and opinions to machines, algorithms, statistics, and rankings. The result is a world where the handprint of the potter, to use an expression dear to Walter Benjamin, clings less to the clay vessel. It feels like the product of an industrial process rather than the work of a craftsman. Is that the way we see the world? Precise, cold, and buffed?

NOT TO MENTION THE NIGHTMARE OF SUPERFICIALITY. That is a real bugbear. There is a deep-rooted suspicion that the perception of the world that has been shaped by new technology is missing a whole segment of reality, perhaps the best part: the one that pulsates under the surface of things, where only patient, laborious, and sophisticated attention will lead. This is a place for which a word was coined in the past that has now become a totem: PROFUNDITY. It gave form to the conviction that—hidden somewhere, at an almost inaccessible depth—things had meaning. It is undeniable that our new techniques for reading the world appear to be specially made to make it impossible to plummet into those depths and almost obligatory to hover over the surface of things with swift, inconclusive movements. What will happen to these human beings who no longer know how to delve down to the roots of things or go to their source? How will their skill at jumping from branch to branch or navigating in the slipstream at full speed be useful to them? Are we all evaporating into a frivolous nothingness? Will this be our last performance?

I haven't written so many question marks all at once for years.

I will be writing about what I think of these fears and other fears like these. There is nothing stupid about them—as some elitist fringes of the revolution would have us believe. They are based on a sum of evidence that, if anything, it would be stupid to ignore. In addition:

▸ inside each of these fears we have sewn a label defining the actions we are taking, which are making us better people. If we manage to answer each of the above questions, we will have an index of our revolution in our hands. A map representing the current situation would, in fact, be a reverse image of our fears. Let us move on then, and unobtrusively cross the border into a new civilization, hiding in the double-bottomed compartment of our doubts the clandestine certainty that somewhere out there is a brilliant Promised Land.◂

The journey is quite exciting, to the point that I often stand by the wayside watching and end up getting delayed, miles behind the people who are actually pushing the revolution forward. From this strange perspective, like an ill-informed, slow-paced but knowledgeable cartographer, I'll be continuing to collect notes and sketches to which I will hazard attaching labels with names and places. In moments of more lucid optimism, I dream of the precision of a map. I dream that every intuition can be put together in the beauty of a globe. Since these intuitions are rare, and I don't want to waste them, writing the book you are reading right now feels like an inevitability—and I will endeavor to do it to the best of my abilities.

Username
▸ Password

Play

Maps

Level Up

PASSWORD

OK, to start with it would be good to know what happened. What *actually* happened.

I'd say the most accredited hypothesis is this: there has been a technological revolution dictated by the advent of digital technology. Over a brief period of time, it has generated an evident mutation in human behavior and in the way minds work. Nobody knows how it will end.

Voilà.

Now let's see if we can do any better.

The term DIGITAL comes from the Latin digitus, or finger; we count on our fingers and that's why DIGITAL means something like NUMERICAL. In this context, the term is used to give a name to a pretty brilliant system for translating any kind of information into numbers, or, more precisely, into a sequence of two figures: either a 1 or a 0. We could equally use a 7 and an 8, but basically what counts is that there are only two figures which correspond, if you like, to *on* and *off* or *yes* and *no*.

So far so good. When I say "translating any kind of information into a list of numbers," I don't mean the kind of information you find in the papers: the daily news, the football results, the murderer's identity. What I mean is any piece of the world that can be broken down into minimal units: sounds, colors, pictures, quantities, temperatures... When I translate a particular piece of the world into a digital language (a certain sequence of 0's and 1's), it becomes incredibly light; it is just a series of numbers. It is weightless and it can exist anywhere; it can travel at lightning speed without spoiling on the way; it doesn't shrink, get dirty, or deteriorate. It goes wherever I send it. Once there is a machine on the other side that is capable of recording the numbers and translating them back into the original information, the game is over.

Take colors, for example. There is no reason why you should know this, but one day a precise numerical value was assigned to every color. If you really want to know, it was decided there were 16,777,216 colors, and a numerical value expressed in a sequence of 0's and 1's was assigned to each and every one of them. No kidding. The purest possible red, for example, is translated into 1111 1111 0000 0000 0000 0000 once it is digitized. Why would anyone do something so poetic? Obviously because, by translating a color into a number, I can put it into a machine that can modify, transport, or just preserve it. It is staggeringly easy, dizzyingly fast, ridiculously cheap, and foolproof. Whenever I want to recall the color "royal crimson," I ask the machine to translate it back again, and it does.

Remarkable.

It is the same with sounds, letters of the alphabet, or the temperature of your body. Pieces of the world.

The trick started to take off at the end of the 1970s. In those days, all the data we stored or transferred was packaged in a system that was called ANALOGICAL. Like everything old, such as grandparents or compasses, analogical technology was a more complete way to record reality. It was more exact, more poetic, but it was also terribly complex, fragile, and perishable. Thermometers with mercury were analogical. Yes, the ones that we used in the old days when we were running a fever. The column of mercury reacted to our body heat by changing volume, and, on the basis of our experience, we deduced our temperature from the way the mercury moved in space. Numbers printed on the glass translated this movement into a verdict: our precise temperature in degrees Fahrenheit (more than 99.5 degrees and you didn't go to school). Nowadays, thermometers are digital. You hold one to your forehead, push a button, and in a second it beeps and tells you your temperature. A sensor records a certain value of a certain temperature that corresponds to a certain sequence of 0's and 1's, which the machine registers, translates into a value in degrees, and writes on the display. The shift in experience could be compared to the transition from playing table soccer to playing video games.

Two worlds.

A thermometer with mercury and a digital thermometer.

Vinyl and CDs.

A movie reel and a DVD.

Table soccer and a video game.

Two worlds.

One possible defect of the digital is that it is not able to record all the nuances of reality. It records reality in leaps and at intervals. Let me explain. The hands of a clock on a church tower are in continuous motion, filling every micro-instant of time. Similarly, as mercury changed volume in the thermometer, it would move up the column filling every micro-level of temperature. Your digital watch does not. It may count seconds for you, even tenths or hundredths of seconds, but at a certain point it stops and ticks to the next number. There, in the gap, is a portion of the world (an infinitesimal portion) that the digital system leaves by the wayside.

On the other hand, the digital system has an invaluable advantage: it is perfect for the computer. That is, for machines that can calculate, modify, or transfer reality provided that it can be processed in the language they know: numbers. This is why, as computers were gradually perfected and slowly became objects for individual consumption, we decided to shift to digital. In practice, we started breaking up reality into infinitesimal particles and attaching to each and every one of these a sequence of 0's and 1's. We digitalized it; we translated it into numbers. In this way, we made the world modifiable, storable, reproducible, and transferable by means of the machines we invented, which were able to perform these tasks quickly, correctly, and cheaply. Nobody noticed, but there must have been one particular day when somebody digitally stored a fragment of the world, and that was the precise fragment that tipped the balance toward digitalization. Don't ask me how, but we know the year that this took place: it was in 2002. Let's use that year as the precise point in time when we went over the peak and found ourselves with the future right in front of us.

2002.

The descent was rapid: the arrival of the web and the often-brilliant application of digital formats to a quite impressive set of technologies generated—with spectacular evidence—what we can now legitimately call a DIGITAL REVOLUTION. It is roughly forty years old, and over the last ten or so, it has officially overthrown whoever was in power before it came along. It is what appears to have dumbed down the new generations.

Fairly straightforward, right? Now comes the hard part.

Revolution is a rather generic word that we use a little too lightly. We can invoke it to describe historic upheavals which piled up mountains of corpses (the French or Russian revolutions), just as we can waste it on such trifles as our favorite team's strategic decision to put three defenders into play (a tactical revolution).

It's confusing.

In any case, a *revolution* takes place when, rather than just inventing a good move, someone changes the whole playing board, causing what is known as a "paradigm shift" (an irresistibly fancy expression that is worth the price of the ticket on its own).

In principle, the expression would seem to fit our case, the case of the digital revolution, very well.

And yet, it is really important to be precise here: there are different kinds of revolution. The revolution that Copernicus triggered when he realized that it was the Earth that revolved around the sun, not vice versa, was not the same kind of revolution as the one we call the French Revolution. Similarly, the invention of democracy in Athens in 500 BC cannot be compared to the invention of the lightbulb (Edison, 1879). All of these people invented new playing boards, but the game does not appear to be the same at all.

When we talk about a digital revolution, for example, it is fairly obvious that we are talking first and foremost about a *technological* revolution: an invention that creates new tools and changes lives, such as the plough, gunpowder, or the railway. Given that we have witnessed many technological revolutions over the years, we have been able to gather some interesting statistics. This is what we have gleaned:

▸ Technological revolutions may be remarkable, but they rarely directly lead to a mental revolution; that is, an equally visible upheaval in the way people think. ◂

GUTENBERG

Example: the invention of the printing press (Gutenberg, Mainz, 1436–40) was a revolutionary change that has always been recognized as having momentous consequences. Leaving most oral culture dead in the water (after it had played a dominant role in a world of illiterates), the press opened up infinite horizons for the freedom and force of human thought. It literally undermined the privilege of disseminating ideas and information that had been controlled by the usual potentates for centuries. All of a sudden, in order to circulate ideas, it was no longer necessary to preside over a network of scribes, which no individual would have been able to afford. Moreover, the machine was faster and simpler, which made it possible to enjoy the fruits of its production. Fantastic.

What is important to point out here, though, is that despite all its miraculous results, the invention of the printing press was fundamentally no more than a dazzling technological acceleration. The earthquake was not on such a scale that it triggered a complete transformation of the human mind. The changes it brought about could be compared to those resulting from scientific or romantic revolutions. Like other technological revolutions, it does not appear to have directly caused a collective mental mutation, as if it had been bogged down before it achieved its aim, giving human beings time to take stock and domesticate its force. It remained a brilliant move in a match that didn't change things, that continued to follow the same rules, and that valued the history of a game that was substantially the same as always.

STEPHENSON

I'll give another example, which is less obvious: the invention of the steam engine (England, 1765). This, again, was not just a brilliant idea: it was world-changing stuff. The Industrial Revolution owes its origins to this invention, and we rightly remember it as a revolution, because it had incalculable consequences, not only on people's daily lives but also, more importantly, on the social geography of the world. The map that once represented the routes money traveled along, and that drew the borders between the rich and the poor, became obsolete the minute the first steam-powered loom was set to work. Everything would change, so radically and violently that most of the blood-soaked skirmish that was the twentieth century could be traced back to the creaking of that seemingly innocuous machine. It's impressive.

And yet, once again, the wave seems to have lapped at the shores of human identity and then withdrawn. Today, if we go back and try to find the crossroads where humanity derailed and moved in new directions, we do not think of Stephenson's steam engine or of the gloomy desperation of the English factories that followed. If anything, we think of Humanism or the Enlightenment: real mental revolutions, which have little to do with technological progress other than a polite nod in its direction. Many centuries later, we can see how these revolutions trickled down, oiling the cogs of the world, lubricating a hydraulic system that was able to shift immense swaths—ideological plates weighing tons—with the ambition of redesigning the structure of human sentiment or the crust of the human planet. They were not just great moves: they were a whole new game.

Simplifying a little, we could in fact say that many revolutions have changed the world—and these have often been technological—but there are very few that have changed people radically. We could call these MENTAL REVOLUTIONS. The curious thing is that, instinctively, WE WOULD PLACE THIS REVOLUTION, THE DIGITAL REVOLUTION, INTO THE LATTER GROUP, AMONG THE MENTAL REVOLUTIONS. Although we see it clearly as a technological revolution, we also see it as being more far-reaching than technological revolutions have usually been; that is, we recognize its capacity to generate a new idea of humanity. This is what we react to, and what triggers our fear. It's not just that we foresee the risks of technological revolutions in general: that many people will lose their jobs, wealth will be distributed unfairly, whole cultures will be wiped out, Planet Earth will suffer, dairies will close, etc. We take note of all of these objections, of course. When the time comes, however, greater fears rise to the surface, regarding the moral, mental, even genetic fabric of society. People predict a radical mutation: a new man springing randomly from an irresistible technological stunt.

This is a crucial point. It requires a certain level of attention, so please put your phones on airplane mode and give the baby a pacifier (since it has yet to be proved that they deform your child's palate).

We instinctively feel that this minor revolution—because it is technological—has the pace of a major revolution of the mind.

It is a signal we should capture in a freeze frame. We should look at it carefully and ask ourselves: what the heck is going on? Are we overestimating what we have set into motion? Are we attributing too much importance to a simple technological leap forward? Is panic overwhelming us? Is the whole thing an unqualified misconception, born from our fears?

It is possible, but I wouldn't bet on it.

On the contrary, I'm convinced that there is something superbly accurate about our suspicion that everything is changing, not just a few things here and there. An astonishing kind of animal instinct compels us to acknowledge that a mutation has taken place; it will not stop at helping us choose a restaurant. Our instinct may be blind, but we can see very clearly.

Where do we go from here?

Let me put it as simply as possible: in all likelihood, we are experiencing a mental revolution. If you are now asking me what happened to the idea that technological revolutions have never before generated this level of chaos, what I have to say is this:

believe me, a trivial error of perspective is confusing the issue. The error is easy to fall into, but it is treacherous and hard to eradicate. WE BELIEVE THE MENTAL REVOLUTION IS AN EFFECT OF THE TECHNOLOGICAL REVOLUTION, BUT WE NEED TO SEE THAT IT'S THE OPPOSITE. We think the digital world is the *cause* of everything, but we should, on the contrary, see it as it probably is. That is, an *effect*: the consequence of some kind of mental revolution. Trust me. Look at the map upside down. Yes, you need to turn it around. You have to invert the darned sequence. *First* the mind revolution, *then* the technological revolution. We think computers generated a new form of intelligence (or stupidity, call it what you want). Invert the sequence right now, and what do you get? A new kind of intelligence that generated computers. What does this mean? It means that a certain mutation of the mind developed, at incredible speed, the tools it needed to live in the world. We can call this a digital revolution. Go on inverting the sequence. Don't stop. Don't ask yourself what kind of mind would lead us to use Google. Ask yourself what kind of mind actually generated a tool like Google. Stop trying to understand whether using a smartphone disconnects us from reality and devote the same amount of time to trying to understand what kind of connection to reality we were looking for when landlines no longer suited our needs. Does multitasking make you feel incapable of focusing? Invert the sequence. What tight spot were we trying to get out of when we invented the devices that finally allowed us to play on more than one board at the same time?

If the digital revolution scares you, invert the sequence and ask yourself what we were escaping from when we opened the door to a revolution of this kind. Look for the intelligence that generated the digital revolution; it's much more important than studying the intelligence the digital revolution generated. It is the original matrix. New humans are not the product of smartphones, they are their inventors; they are the ones who felt the need for devices of that kind and designed them for their own use and consumption. They constructed them to escape from a prison, answer a question, or stifle their fear. Pause.

One last thing.

- All the digital strongholds that mark our landscape should thus be seen as geological formations pushing upward and outward as a result of an underground earthquake.

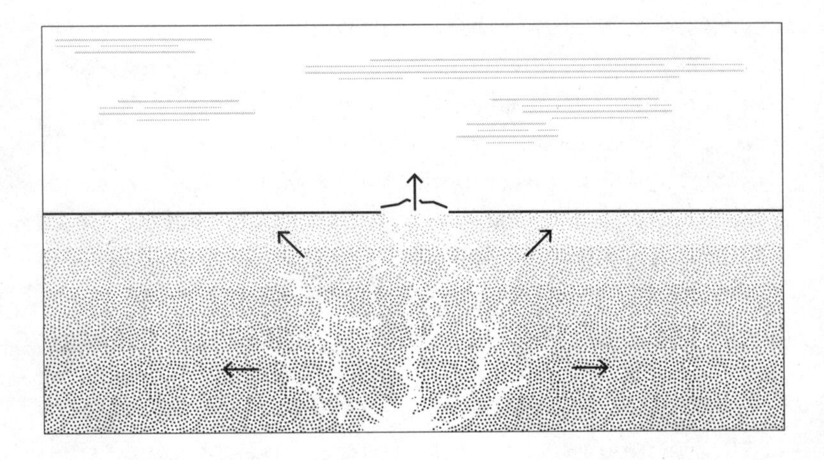

The earthquake was the mental revolution that gave birth to us. It happened somewhere else, in a time warp that we are hardly aware of and know little about, but we can recognize and study the spectacular alterations in our actions, habits, and mental postures that it produced on the earth's crust.

Many of these alterations can be traced back to the mental revolution. These alterations—more than others—appear to us like a scripture preserving the last codes of the mutation. As long as we don't hold them responsible for causing everything, they have a great deal to reveal and to teach us.

We should see them as ruins: archeological remains from which we can reconstruct the wonders of a lost civilization.

Our civilization. ◂

You can put your mobile data back on now, thank you.

Ah, and the baby is crying.

In short: turn the darned map around.

The digital revolution is underneath, not on top.

That's right.

Get used to thinking of the digital world as an effect, not a cause. Shift your gaze to the place where it all started. Look for the mental revolution it all stemmed from. If there's a model of humanity this is all working toward, it is written there.

OK.

The map may still be almost completely blank, but at least we're holding it up the right way.

Believe me, that was the most difficult part.

Now we can start measuring, adding a name here and there, drawing a few boundaries.

Moving on with the idea of the digital revolution as a range of mountains produced by an underground earthquake. Now let's try to draw it.

Username

Password

▸ Play 1978 Vertebra zero

1981–98 *From Commodore 64 to Google:*
The Classical Era

Commentaries on the Classical Era

1999–2007 *From Napster to the iPhone:*
Colonization

Commentaries on the Colonization Era

2008–16 *From Apps to AlphaGo:*
The game

Commentaries on the Game Era

Maps

Level Up

1978. VERTEBRA ZERO

Even though the digital revolution is a constellation of a diverse and articulated array of phenomena and events, an underlying structure—something like a backbone or a mountain range—can be detected. What we are trying to understand here is the way the peaks of different altitudes—geological formations thrust up by seismic activity—are aligned. Let's start by isolating what we might symbolically call VERTEBRA ZERO. I don't want you to imagine anything particularly serious here; what I have in mind is a video game.

It was called *Space Invaders*. Millennials probably don't even know what it is. I do, because I used to play it. I was twenty years old, and—weirdly enough—I had a lot of time on my hands. A Japanese engineer called Nishikado Tomohiro invented it. The game consisted of shooting at aliens who dropped out of the sky in an idiotic and repetitive, but terrifying, fashion. They speeded up as they descended and, once they were on top of you, you didn't know what had hit you.

Compared to today, the graphic design was primitive. The aliens (we called them "little Martians" in Italy) were miniature, black and white, two-dimensional spiders that looked like they'd been drawn by an imbecile. Death announcements in the newspapers were more interesting.

There weren't any home computers at the time. To play *Space Invaders* you had to go to public places, like cafés, where there was a cabinet-like box—it's hard to imagine nowadays how big it was—with a screen the size of a small television and a basic console with three keys or, in the more sophisticated versions, a joystick and two keys.

You had to bend down, stick a coin into a slot, and press *play*; that was when you started wildly pushing the keys, shooting little Martians like crazy. In Japan, you needed a 100 yen coin. There were so many people playing *Space Invaders* that the coin soon went out of circulation, and the country had to mint a whole lot more in a hurry.

The success of the game can teach us a great deal, especially if we go back even further and recover the memory of the two games that were popular in cafés before the *Space Invaders* cenotaph came along: table soccer and pinball.

Which brings us to the point.

If you take a step back into the past—well, maybe two—you can reconstruct a sequence of games that, more than anything else in the world, allows you to FEEL, rather than fathom, what it was like before the digital revolution.

The sequence is this: table soccer, pinball, *Space Invaders*.

Don't make that face. Trust me.

Think carefully about this sequence: try to feel it physically, go back and play each of those games in your mind, one after the other. Your body will realize that, with every shift, something melts away. Everything becomes faster, more abstract, ethereal, fluid, artificial, and abridged. A mutation. Very similar to the mutation that led us from the analogical to the digital.

It's not so much in the brain; it's really very physical. Playing table soccer, you feel the thrust in the palm of your hand; the sounds are naturally produced by the mechanics of the game; everything is real; the ball actually exists; you get physically tired as you move around; you sweat. With pinball, there's already a shift. The game takes place under a sheet of glass; most of the sounds are electric and artificially reproduced; the distance between you and the ball is greater; everything is reduced to two keys which are only distantly related to the ball and give you no more than a semi-perception of what is really happening. While with table soccer your hands express themselves in an infinite range of speeds and positions, with pinball everything is down to two fingers. While these retain a limited range of options, they are generally exploited only by the most expert players. As far as the body is concerned, with pinball it is a silent witness, almost entirely excluded from the game. The last vestige of body participation is a movement of the pelvis with a rather crude sexual overtone that was used to deviate the course of the pinball. For both of these reasons, an over-accentuated use of the pelvic thrust is not recommended.

Now try playing *Space Invaders*.

Your body? It's simply not there. There's almost nothing physical (in the sense of tangible) in the game: the equivalent of the ball—the little Martian—is not real; neither are the sounds. The screen, which didn't exist in table soccer and which was only used in pinball to keep track of points, has taken over everything else and has BECOME the game itself. Everything is immaterial, graphically represented, and indirect. If there is such a thing as reality, it is a representation behind a glass screen that you cannot modify except by means of external commands that impersonally communicate your orders. This description makes it sound cold, coercive, claustrophobic—basically, sad. However, try playing it and you'll suddenly feel the lack of attrition; the smoothness of the game board; the lightness of your touch; the almost fluid flow of commands and decisions; the concentration of every possible game strategy into the very essence of gaming; the purity of the system; the potential for almost absolute focus; the speed with which things happen. I bet you'll soon realize why they ran out of coins so quickly in Japan.

Now flash back in a nanosecond to the table soccer knobs. It's a shock, right? It's as if they airlifted you out of a meditation session right into the middle of a heated argument in a bar. Everything feels chunky, clunky, imprecise, and annoyingly real... It's not that one is better than the other, there's no way of saying it, but they are certainly different. Very different. Which one do you feel most yourself with, most present, most alive?

Play table soccer for a bit and then return quickly to the *Space Invaders* console.

Go back and forth between the two, maybe stopping every now and again at the pinball half-way station.

Do it for real.

Can you feel the migration?

I say MIGRATION for a reason. Shifting the center of gravity, sliding many details from one part of the landscape to another, changing the positioning of your skills, your potential, your sensations, your emotions. CHANGING THE SUBSTANCE OF YOUR EXPERIENCE.

They are just three games, but there are so many things that migrate during the journey from the older to the newer experiences.

Don't waste time attempting to decide which is better and which is worse. Concentrate on trying to narrow down your perception of the migration into an overview, a single sensation. Above all, a sensation.

Did you succeed? Good. The sensation you are channeling is similar to the flow of the analogical to the digital. You are touching the central nerve of the revolution. Its basic movement. Its secret, if you will.

Space Invaders, with its unpretentious aim of being a game for people with nothing better to do, is one of the first geological traces of an earthquake.

The core of the game, of course, was already entirely digital—a software contained in a disc. If we can talk about the digital revolution having a backbone, then this core would be the first vertebra. Under the skin of the world it only protrudes very slightly, but it can be felt and it can be seen. It exists. It is a beginning.

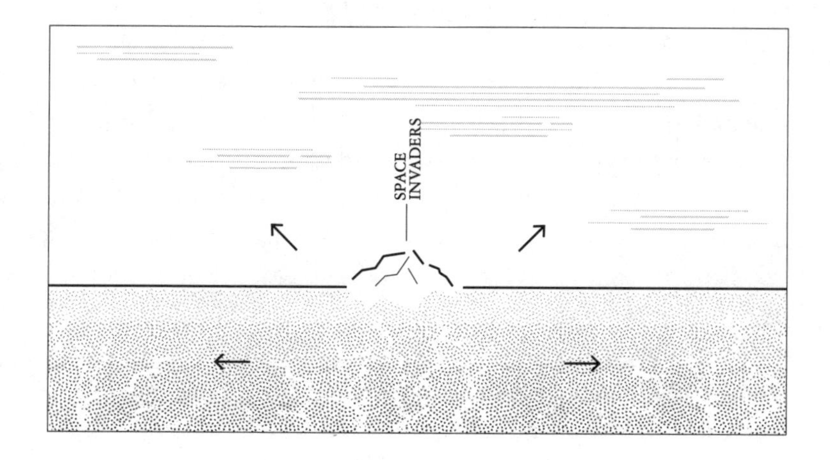

NOTE

To give you an idea of the task that awaits us, I'll pause for a second and look at this vertebra—as we will be looking at all the others in the backbone of the digital revolution—as if it were an archeological site that provides us with clues to a lost civilization. We need to look for fossil traces of previous forms of life. The codes of the mental revolution that generated everything that followed.

It's quicker to do it than to explain it. So, I'll just brush the dirt off the first vertebra and bring to light a few clues.

FIRST

Compared to table soccer and pinball, *Space Invaders* was a game that established a revolutionary shift in physical and mental posture. To put it in an incredibly condensed fashion, brutally boiled down to the basics, this was: human, console, screen; or human, keys, screen. Fingers on keys, eyes on screen. Commands given with fingers on keys, verified by eyes on screen. Add a little audio to make the system more functional. Does this remind you of anything? It is currently one of the physical and mental postures we spend most of our time engaged in. We adopt this posture for executing operations of every kind, whether it is booking a hotel room or telling someone you love them. It is the posture that defines the digital era par excellence. Not even the advent of touch screens managed to alter it that much.

It wasn't *Space Invaders* that triggered it, of course. However, it is highly likely that the posture occurred for the first time while playing the game, when it emerged from the surface of our existence in a statistically significant number of people. To give you an idea, the first personal computer that enjoyed a certain popularity (though nowhere near as popular as *Space Invaders* and other arcade games) was the Commodore 64 that came out in 1982. The first Mac—which is to *Space Invaders* the equivalent of a cathedral compared to a votive chapel—came out in 1984. It was another twenty-one years before the first smartphone went on the market in 2003.

So, if you rewind a little and look back at the first time that posture—human-keys-screen all rolled into one—crashed into the everyday life of an awful lot of people, what do you think you'll see? *Space Invaders*, I believe. And other games of that ilk.

WHAT CAN WE LEARN?

That, as it happens, in the DNA of the vertebra zero there is a posture that would go on to have a glorious future, and that we would recognize in the geological formations that we are calling the technological revolution: human-keys-screen. It is the posture I'm writing this book in (but probably not the one you are reading it in: honor to the old-fashioned book that still resists any mutation).

SECOND

Table football was a piece of furniture with a sense of dignity; the pinball machine was beautiful in its own way. The cabinet containing *Space Invaders* was truly ugly. On the other hand, table football could really only be table football; the only thing you could do to alter it was to change the color of the players' shirts. Pinball machines came in a whole range of different designs (from fantasy worlds to soft porn with semi-naked women). There was also the possibility to make the circuit more challenging with obstacles and bridges, but basically the principle was always the same: the ball was bounced into position and then had to find its way home. That was it. The horrendous *Space Invaders* catafalque, by contrast, CONTAINED INFINITY. Once the posture human-keys-screen had been established, there were no limits. There were all the games in the world in there; all you had to do was change the disc. For those who could see ahead, there was already the potential for *FIFA 2018* or *Call of Duty*. All that was needed was a tweak in graphic design, a few extra functions, and more advanced video and audio technology. Fifteen years or so later, this would be achieved to spectacular effect with the 1994 PlayStation.

WHAT CAN WE LEARN?

That in the DNA of the vertebra zero there is a type of movement that would go on to have a glorious future and that we would recognize in the geological formations that we are calling the technological revolution: rather than generating many different, attractive worlds, all its energy was invested in inventing a single ambient in which all existing worlds can be contained. In other words, there's no point in wasting time designing things that cannot be developed extensively; try to come up with an invention that has infinite possibilities because it was designed to contain EVERYTHING.

THIRD

Space Invaders was a GAME. I don't know if you can discern the delightful implications of this concept. In practical terms, what we are talking about is underground seismic activity that managed to break through the habitual crust of daily life on earth. The beauty of it is that the first time this happened—or, at least, one of the first times this happened—was at that moment in their lives when humans had put their slippers on, put everything else on hold, and decided to play a game. I find this idea heartwarming. I wonder if it happened by chance. Of course, I'd like to think that the day we decided to turn the tables—and throw ourselves into starting a revolution that would change the world—was a day of vacation. Barefoot and downing a can of beer.

WHAT CAN WE LEARN?

That in the DNA of the vertebra zero, there was an attitude that would go on to have a glorious future and that we would recognize in the geological formations that we are calling the technological revolution. That is, in order to generate change, we need to create tools that are also games, or at least look like them. We are frivolous divinities: we like to create things on the seventh day, when the real gods are resting.

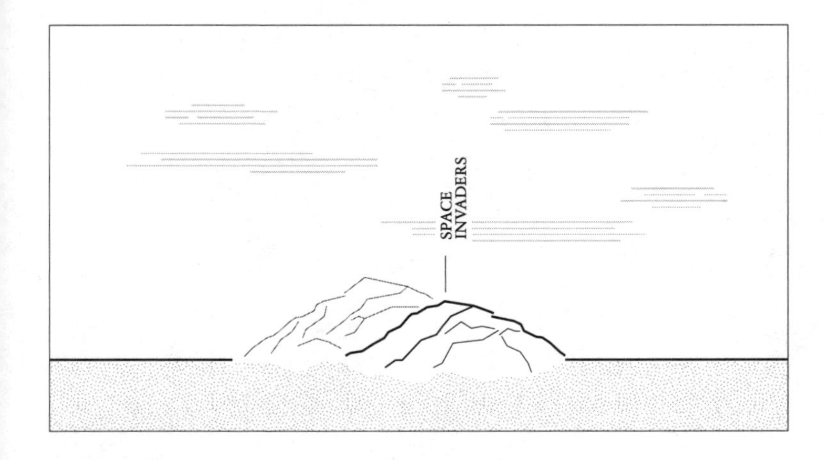

OK, I'll stop here for now. You must be beginning to see that, if we can deduce so much from a single, tiny vertebra, the idea of studying the rest of the backbone is irresistible.

It is thus worthwhile continuing. Next chapter: next section of the backbone; next archeological site to explore. I'm starting to have fun, for real.

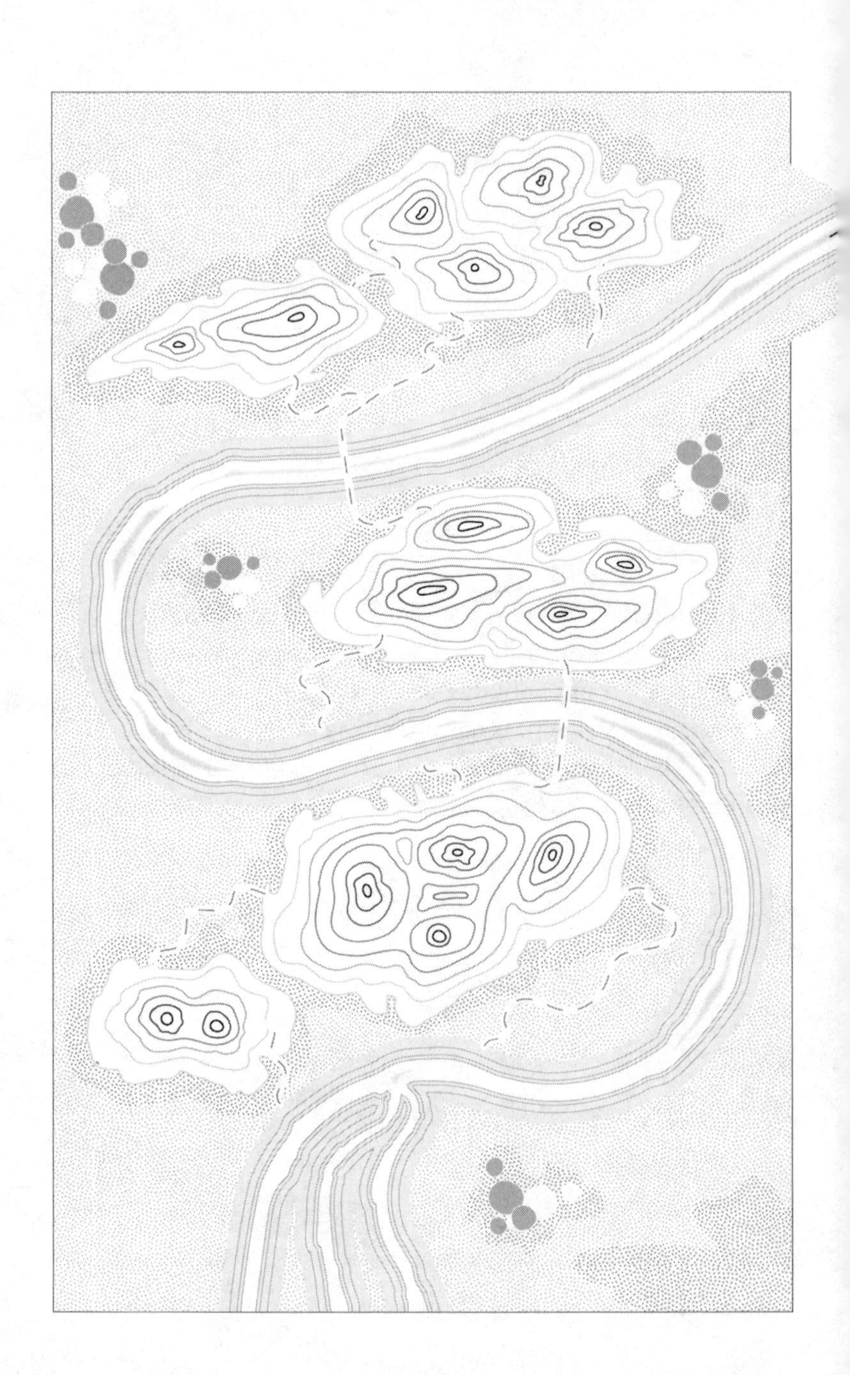

1981–98.
FROM COMMODORE 64 TO GOOGLE.

THE CLASSICAL ERA.

Almost twenty years to adjust the game table.

An inevitable but important premise is in order. If we want to reorganize the entire galaxy of events that we are calling a DIGITAL REVOLUTION into a coherent backbone structure or into a well-articulated mountain range, we are obliged to gloss over the nuances and present an abridged version. What we are doing is recording only the peaks, even if this means sacrificing the details of the processes that may well have gone on for decades. In this book we have decided to record only the events that have been thrust up from under the earth's crust as instances of collective consumption, not only among certain elite minorities but also by the masses. I know the method is random, but at the end of the day we need a legible outline of events so badly that we cannot be held back by the cult of precision. What I suggest is that you enjoy the chance to take a look at everything from above—as if it were an aerial photo—and that for a few chapters you suspend your judgment concerning the unavoidable mistakes created by such a bird's-eye view. Whenever possible, we will swoop down and look at things closer up. I promise.

Let's set *Space Invaders* aside, shall we? What we are looking at in this chapter are the first real mountains in the range. These started emerging in the early 1980s.

1981–84

- After many years of experimentation, within four years of one another, three different personal computers were brought out onto the market and sold so successfully that it was possible to own one even if you were not a member of the elite, a genius, or a professor at Stanford: IBM's model, Commodore 64, and Apple's prize product, the Mac. If you look back at the early prototypes, they seem terribly clunky, but back then they must have appeared quite elegant and passably user-friendly. Of the three, the Mac was the least commercially successful, but the most inspiring. It was the first to organize the material graphically, with the aim of being idiot-proof. There was a desktop, you opened up windows, you threw stuff in the trash: these were all actions people could recognize. You moved around on the screen by gyrating a strange object called a *mouse*. It's easy to understand how the equation between intelligence and boredom began to lose its meaning.

ZOOM

It's hard to appreciate the significance of all this if we don't pause for a moment and think about the P in the expression PC.

Personal.

Nowadays, the fact that almost everyone has a computer at home is totally normal. You mustn't forget, however, that only forty years ago, it would have sounded like madness. Computers had been around for years, but they were gigantic monsters that processed data in the laboratories of a few institutions whose primary aim was mostly domination or some form of supremacy. The idea that they may one day end up on all of our desks would have been considered truly visionary at the time. I'll even go as far as to say that the real act of genius was not so much inventing computers, but imagining that they may one day become an everyday tool for personal, individual use. Smoldering below the surface of that vision was the revolutionary desire to give any individual the power that had been created for the few. Incredible. This is why, when you look at a Commodore 64, rather than ask yourself why they opted for that sickly color, you should appreciate that the world was TRULY turning at that moment. Not a second before.

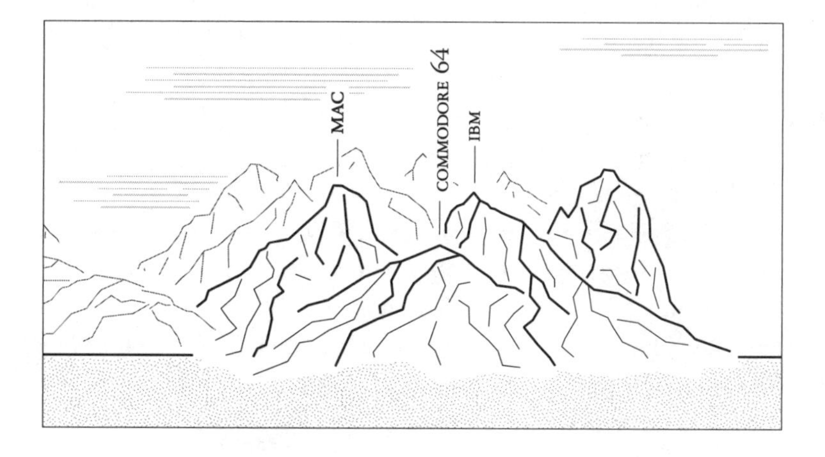

• In 1981, the first email protocol, SMTP, was published. The system was so easy that the circulation of emails increased at a dizzying speed (thirty years later, in 2012, human beings were corresponding at the rate of 144 billion emails a day, three-quarters of which turned out to be spam). The very first email, for your information, had been sent many years before, in 1971, by a thirty-year-old American called Roy Tomlinson, who was studying to be an engineer in New York. The @ symbol was his idea, I discovered.

IMPORTANT

Emails traveled from one computer to another by means of something we might call an invisible network that most normal people at the time had no idea existed. Those in the know called it *The Internet*. Imagine it as a subterranean St. Barbara, the patron saint of fire fighters. If you persist for a few more lines, you will appreciate the immensity of the explosion that would, in the space of a few years, split the earth's crust and thrust up into the atmosphere one of the highest peaks ever produced by the digital revolution.

1982

- Rising up above the earth's crust, the wave of digitalization that would wash over the world was impossible to hide. The first sign was the appearance of the first music CD: a recording translated into digital form and stored on a compact disc the size of a saucer. Philips and Sony—the Netherlands and Japan, respectively—launched the innovation together. Their first music CD was an inexplicably third-rate piece: Richard Strauss's *Alpine Symphony*. (Mind you, the first CD of pop music was by ABBA.)

1988

- Another important step in the progressive digitalization of the world after music: images. The first completely digital camera was produced by Fuji, a Japanese company, of course.

DECEMBER 1990

- A British IT engineer, Tim Berners-Lee, inaugurated the World Wide Web, changing the world forever.

It is, as we all know, a historical moment. A good half of the world as we know it today was born at that very moment. I will still be claiming this even if the day after tomorrow the web is swept away and replaced by something better (which is happening, by the way). In the invention of the web, there was a mental act that would soon become a habitual move for the brains of billions of human beings. Together with a couple of other stupefying moves, this laid the foundation of our new civilization. So, concentrate. We need a serious parenthesis here; this is where we need to understand things properly. At least, I needed to.

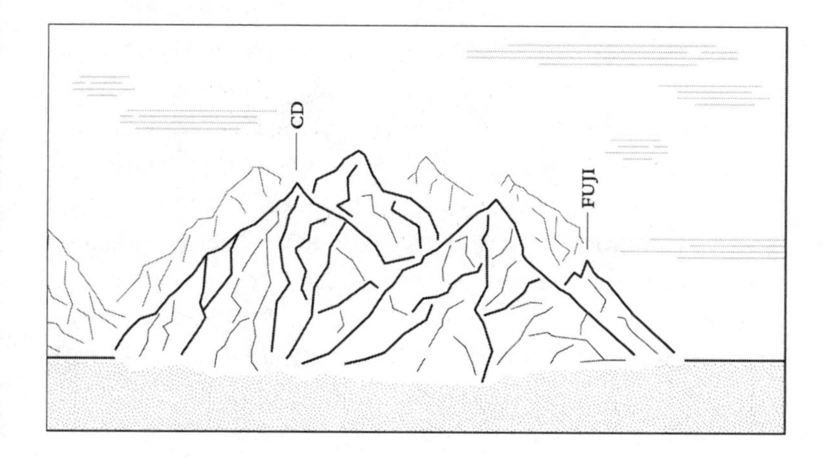

I think it would be useful to start with a fact you may not like that much: The internet and the web are two different things. I know, it's annoying, but just accept it for now. The internet came about before the web, long before. Let me try to explain how it all happened.

It all started in the Cold War years as a result of the US Military's paranoid need to be able to communicate without the Soviets poking their nose into their business. They worked on a solution and came up with a clever idea in the 1960s that they called ARPANET: a system whereby physically distant computers could communicate with one another by means of a previously nonexistent method for packaging data, thus creating a kind of secure circuit for computers to pass on information without any communists being able to get their hands on it. The communication, it must be added, took place in a ridiculously short amount of time. A key was pressed, and the message got to the other side of the world in a moment. Well, if not in a moment, amazingly quickly.

If they had not been in the thrall of their obsession with the communist threat, they might have realized sooner that an idea of this kind opened up incredible new horizons—well beyond the military sphere. A few American universities that had worked on developing ARPANET took notice. They honed the technology and used it to build a communication circuit for their computers. On October 29, 1969, a message sent from a UCLA computer arrived at Stanford University, traveling 350 miles in a flash. Only half the message got there, but the glitch was soon rectified, and Stanford received the second message in its entirety. The researchers went on to create a grid connecting all their computers. To start with, they used it to send letters (which we now call emails), but as time went on, they were able to share their research papers, whole books, I imagine even jokes. Whatever it was, it was quite impressive.

At that point, many other universities, some big corporations, and a few nation states began to grasp the incredible potential of the technology and built circuits of their own, so that all their computers could communicate with one another. What they were creating was a *network*. Each entity had its own, and each network had its own set of rules, mechanisms, and functions; in short, they were all speaking different languages. Nothing would have happened—and you'd still be licking stamps—if in 1974 two American IT engineers hadn't invented a protocol that allowed all existing networks to connect with one another and magically communicate. It was like an instant planetary interpreter: every network spoke its own language, and the protocol translated it on the spot. Being engineers, they didn't give it a very nice name, but it is worth learning nevertheless: TCP/IP. It was the invention that swept away the barriers between all the different networks, achieving the formidable result of creating a single world-wide network that some people called *The Internet*.

This was in the early 1970s, and it's important to remember that these networks interested a ridiculously tiny percentage of the world's population. It was the same elite that had access to computers. It was a niche game. I think there must be more people playing the game of curling today. This is why there is no sign of this breakthrough registered on our imaginary backbone of the digital revolution, since—as I have explained—only those seismic events that actually changed people's lives appear above the surface. In this story, it was not until 1990 that these events began to emerge. Tim Berners-Lee, who worked at CERN in Geneva, invented what he called the web. (This is the first time that old Europe gets a look in; up until now all the heroes have been American, most of them Californian. Full disclosure is in order, however: Berners-Lee's breakthrough took place using an American computer called NEXT, produced by a Californian company that was founded—significantly—by Steve Jobs).

What did Berners-Lee invent, then? Not the internet, it is now clear. So, what breakthrough did the web represent? I've learned that there are many possible answers to this important question, and that they are all hopelessly indefinite and incomplete. I'll add one of my own.

Whatever the web is, Berners-Lee invented it in three clear moves.

The first was born from a question: if all the computers in the world can communicate on the internet, why should we be content with so little? Let me explain. Imagine the computer on Professor Berners-Lee's desk, and then try to picture his study. Look around:

there are almost certainly some cupboards and filing cabinets. Open the drawers, the many drawers, the hundreds of drawers, and you will see they are full of papers, projects, ideas, notes, vacation snaps, love letters, prescriptions, Beatles CDs, collections of Marvel comics, old bank statements, film club membership cards, and so on. Now ask yourself: why not dive straight into those drawers? How come information can travel thousands of miles (thousands!) and then, two yards (two yards!) from everything that's in the drawers, I can't get in and I have to stop at Professor Berners-Lee's computer? I can't delve into the drawers and sort through everything they contain. It's plain stupid. So, I have a chat with Professor Berners-Lee. He's a good listener and, since he knows what to do, he invents a system that modifies the structure of the drawers, so that I can walk those extra two yards and take a look inside. Of course, he doesn't have to open all of them; he can choose which ones to make available. When he has done so, he applies himself to giving them a structure that allows me to reach them, open them, rummage through the contents, and even take anything that interests me away. How does he manage this? By duplicating the contents of the drawers into a myriad of digital representations, which he then puts in a place that he quite naturally calls a *site*. Website. He sees it as a tree with branches that reach up into space where every leaf is a page—a web page. What is the tree made of? Digital representations. That is, texts, images, and sounds that are stored on the computer after having been formatted into a digital language. Once they are there, the infinite network of the internet becomes available. By using that network, the contents of Professor Berners-Lee's drawers, duplicated into a digital version, can start their journey. They can reach my desk, where, at the end of the process, I can find what I wanted all along: Professor Berners-Lee's collection of Marvel comics (I was less interested in his medical prescriptions).

Remarkable, huh?

But a little predictable, if it were not for the fact that Professor Berners-Lee went one move further, which was *truly* exciting. In order to simplify things, and make them more spectacular, he PUTS ALL THE DRAWERS IN COMMUNICATION WITH ONE ANOTHER. This means that when I open one drawer, I can go on and open another without even closing the first and without having to go back to square one. I can do this thanks to a little door that professor Berners-Lee designed and called a *link*. These are words with a special function, more than words, they are *hyper-words*, that appear in blue. I can click on them and end up inside another drawer. As you can appreciate, this is starting to be really fun. A mere hour or two ago, sending an email felt like something out of the ordinary; now that I can flit from one drawer to another like a butterfly, restricting myself to sending boring emails feels like senseless self-punishment. It's so much more exciting to flutter from drawer to drawer, from one website to another. Especially after Professor Berners-Lee made things even more exhilarating by developing the third move.

Rather than keep the breakthrough to himself or attempt to put it on the market, the professor (with the permission of his employer, CERN) decided to make his system for opening drawers public. His idea was very simple: if we all connected all our drawers by means of these links, we would find ourselves with a formidable web of information. Anyone could navigate this web, whenever they wanted, taking whatever they needed: it would effectively be a *World Wide Web*, a spider's web as big as the world, open to everyone, where every single document in the world—every text, photo, sound, video—would become readily available. Add to this one last, irresistible factor: it would be free.

Wow!

What's not to like?

Nothing, of course, and that's why we're here.

In 1991, there was only one website: Berners-Lee's.

The following year, nine more opened up.

In 1993, there were 130.

In 1994—2,738.

In 1995—23,500.

In 1996—257,601.

Today, as I write this, there are over a billion.

As you must understand by now, the effects of this tsunami have been dramatic. We are most interested in the mental effects, of course. You will find an examination of a ridge, this mountain that suddenly appeared, breaking through the crust of people's habits, and went on growing precipitously every year (to give you an idea, in the time it took me to write these few lines, at least a thousand new websites have been opened. How do I know? Check out www.internetlivestats.com).

As I was saying, we'll look at the effects on people's minds in a little while. If we can set aside this gargantuan phenomenon with a vague sense that we've understood something about how it happened, it's already a good result. Do you have that sense? I hope so. Let's go back to the backbone, then. We had gotten to 1990.

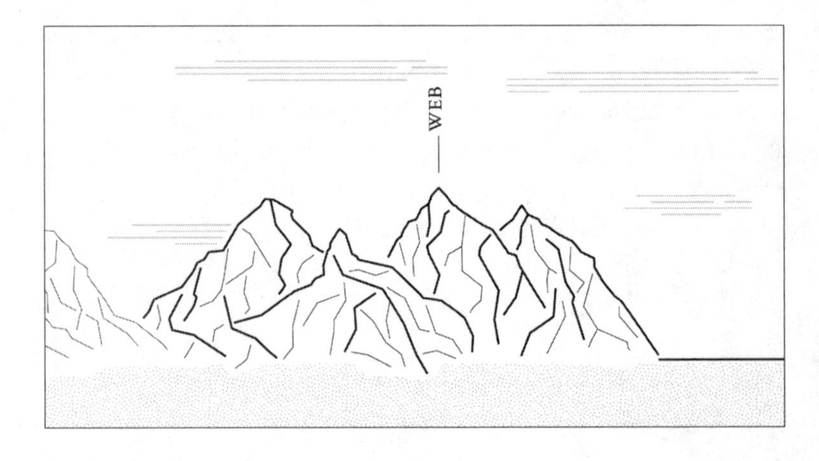

1990

- Tim Berners-Lee launched the World Wide Web and changed the world.

1991–92

- Nothing special that I know of happened. Maybe people were still dealing with the aftershock.

1993

- A group of European researchers invented MP3, a system for making audio files lighter, reducing their digital weight even further than before. The concept of COMPRESSION was invented, which would later be applied to photographic images (JPEGs) and video images (MPEGs). The concept was that if you manage to strip a digital sound file of all the numerical sequences that are not strictly necessary (for example, background noises that are inaudible to the human ear), you end up with poorer sound quality, but the file is much lighter and therefore easier to export, send, or store. There's no way you would be able to listen to music on your cell phone without this level of compression. (Needless to say, CDs immediately started to feel like artifacts from a bygone civilization.)

- Mosaic was launched. It was the most-used early browser for surfing the web. Pivotal. Berners-Lee had invented a parallel world that was digital (the web) but he had not provided any access service at all. Before Mosaic came along, in order to navigate the web, you had to be a computer whiz or like Indiana Jones in explorer mode. A browser provides a set of services that allows an idiot like me to explore the meanders of the web. I install it onto my computer, and it lets me navigate without really having to know how it's being done. (A little like cruise ships take you around nowadays without really knowing where you are going.) Mosaic was the first browser

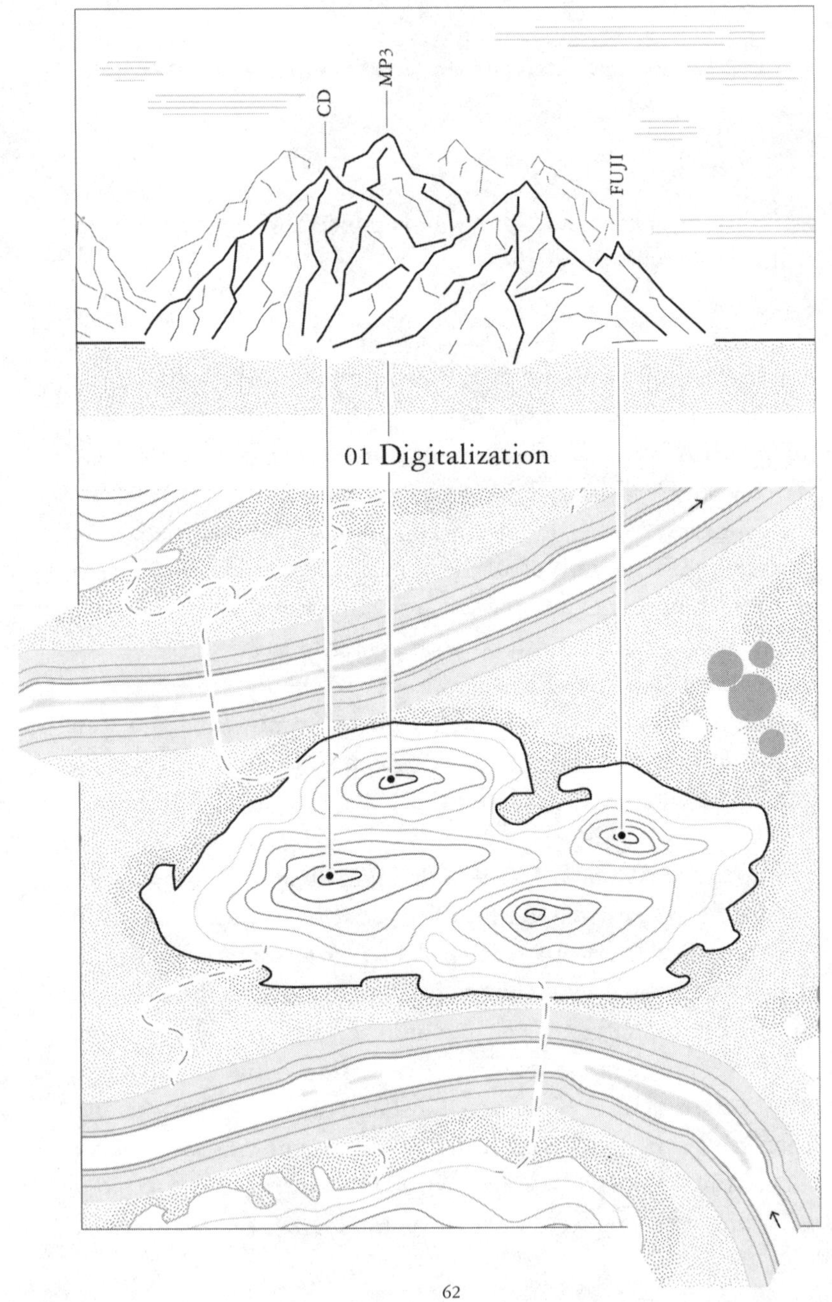
CD
MP3
FUJI
01 Digitalization

that was vaguely successful. It was designed by two students from the University of Urbana-Champagne, Illinois. Today, it no longer exists. However, browsers are still essential. Nowadays, they are called things like Safari, Google Chrome, and Internet Explorer. Without these browsers, the web would still be a place for a few nerdy engineers with time to waste.

1994

• Cadabra was born in Seattle. The name means nothing to you, but it should, because it was Amazon's first name. The idea was to create an online bookshop where you could buy every book in the world. Basically, you could switch on your computer, choose a book, pay for it, and take it home without ever lifting your butt from your chair. It was a crazy idea, but the guy who had it evidently placed great trust in a number that's worth noting: the year before, recorded internet users increased by 2300 percent. As well as changing the company's name a year later, the founder—Jeff Bezos—soon realized that limiting his business to books was stupid. Nowadays, you can buy a car on Amazon. Or a hair dryer.

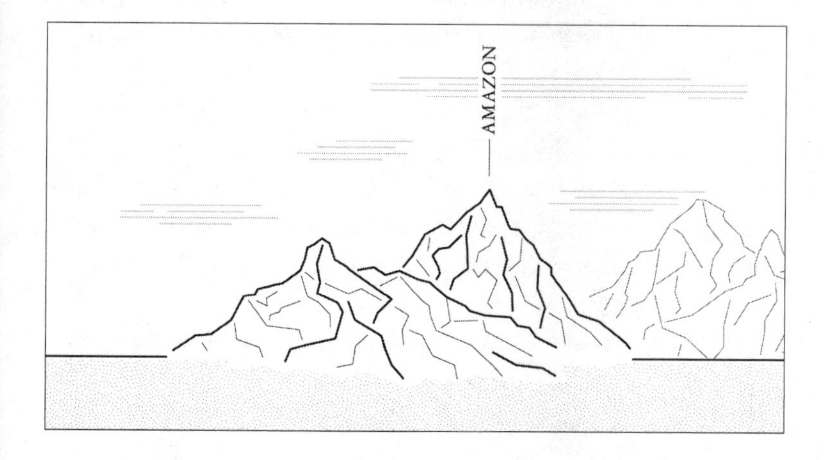

Another pause is required here, because it's important to remember the way things went. History has it that when Jeff Bezos needed to raise cash to fund the first couple years of Amazon, he asked his father, among others. He wanted to convince Bezos Sr. to invest his life savings of about $300,000. He had to painstakingly explain everything, and I'm sure he did so convincingly. Bezos Sr. listened for a while and eventually asked his son the following question: "What's the internet?"

The question may seem comical today but it will help us focus on what things were like in those years. This is what the pause is for: to focus on those years.

As a personal example, in those years I was in Santa Monica, California, spending the first money I'd ever earned by allowing myself the luxury of sitting in a hotel room writing a play that turned out, to my great surprise, to be total crap. Every now and again, I would go out and stretch my legs on the promenade and, one day, I made my way into a bookshop. I must have been checking out all the book jackets and appreciating the evident superiority of American designers to ours when I bumped into—and I remember this as clearly as if it were yesterday—a kind of book I couldn't understand. I had no idea what it was for, although I did remember something a friend had once told me. The warning sign was the fact that the book looked like a catalogue of places, names, or titles (I wasn't sure which). They all had a dot in the middle, two backslashes, and acronyms like CH and EU, but it's hard to remember exactly. In short, they all looked alike, but none of them looked like anything I'd ever seen before. Now I know what they were: websites. Now I know that the book I had in my hands was a kind of phone directory—like the yellow pages—for website addresses. The fact that they were selling it in such a cool bookshop in Santa Monica tells you a great deal about the newborn status of the digital revolution. They must have had absolutely no idea where they were going if they printed directories of websites *on paper*. There they were, in alphabetical order and poignantly divided into categories: lists of websites about sports, gastronomy, or doctors, for example. It's heart-wrenching, isn't it? It was a bit like when car engines were calculated according to how many horses might have been able to shift the same weight. These are the moments when a new age is dawning, but human genius is forced to cohabit with a form of irremediable, imbecile hesitation. Moments when, even if you happen to be Jeff Bezos's father, you can still ask, "What's the internet?" without being considered an idiot. Anyway, I bought the book thinking I'd give it to my friend, just as I might buy a book of Japanese grammar for another eccentric friend studying a subject that I considered completely irrelevant. I myself had no idea what a website was at the time. I didn't know in the most radical, definitive, and shameful sense of not knowing. That is, I had no inkling whatsoever what kind of thing it was, nor what form or identity it might possess. The web was simply not on the list of the things that I knew, although that was the least of it. The logic of the web, its form, its mental architecture didn't exist in my mind. Not only did I not know it existed; I didn't possess the mental categories that had generated it.

I'd like to point out that I was a college graduate. I had majored in philosophy. What I'm trying to say is that it wasn't a personal failure. We were all ignorant, not just me and Bezos Sr.

Now that we've glimpsed the backbone of the digital revolution, let's try to feel the vertebrae one by one with our fingers. In those years, they were still tender, provisional, mutable cartilage. They were new organisms, both in their conception and in their structure. They were alien material.

My friend now writes books, good ones. Bezos Sr. doesn't, but he did give his son Jeff his nest egg of $300,000, and I imagine it made him a fair profit.

OK. Let's return to the backbone. We'd got to 1994, when Amazon was launched, but that was not the only noteworthy event of that year.

1994

• IBM produced its first smartphone. Cell phones had been around for a while, but this was the first phone that did things phones were not supposed to do, like allow you to send emails and play video games, for example. The product survived six months before it was taken off the market. False start. Another nine years were to go by before smartphones for mass consumption emerged. I don't really know why.

• The PlayStation was born. It was produced by the Japanese company Sony. The relationship of parents with their children would never be the same again. Neither would our relationship with reality, as we shall soon see.

• Yahoo! was also invented, launching the fashion for stupid names. It was, anyway, a historic moment. The portal was designed by two students at Stanford, who decided to do something totally obvious: eliminate that pathetic paper directory—the yellow pages of the web—the one I'd given to my friend. Finally, someone could help you navigate the web and find your way through the different sites on the internet. It wasn't rocket science.

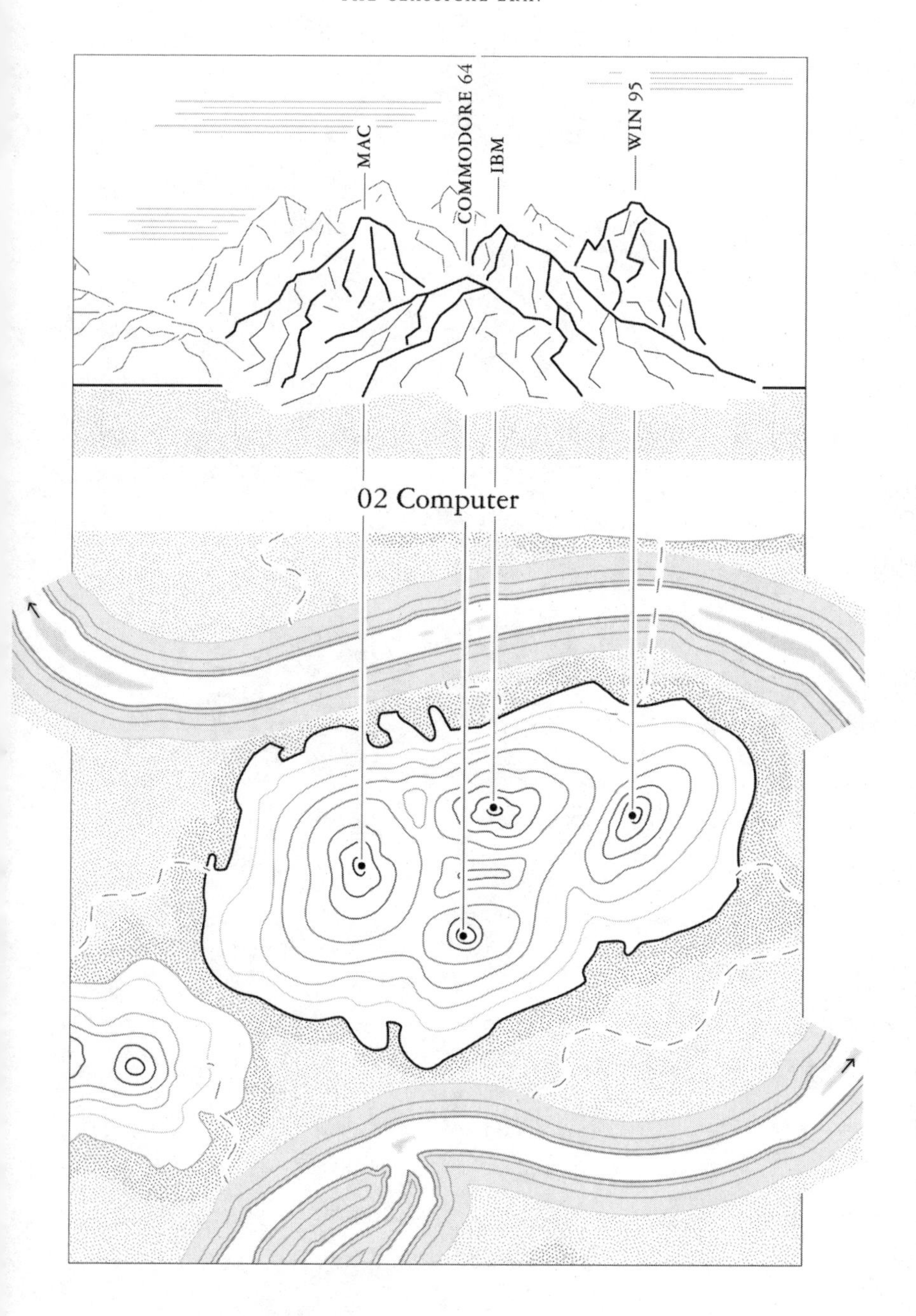
MAC
COMMODORE 64
IBM
WIN 95
02 Computer

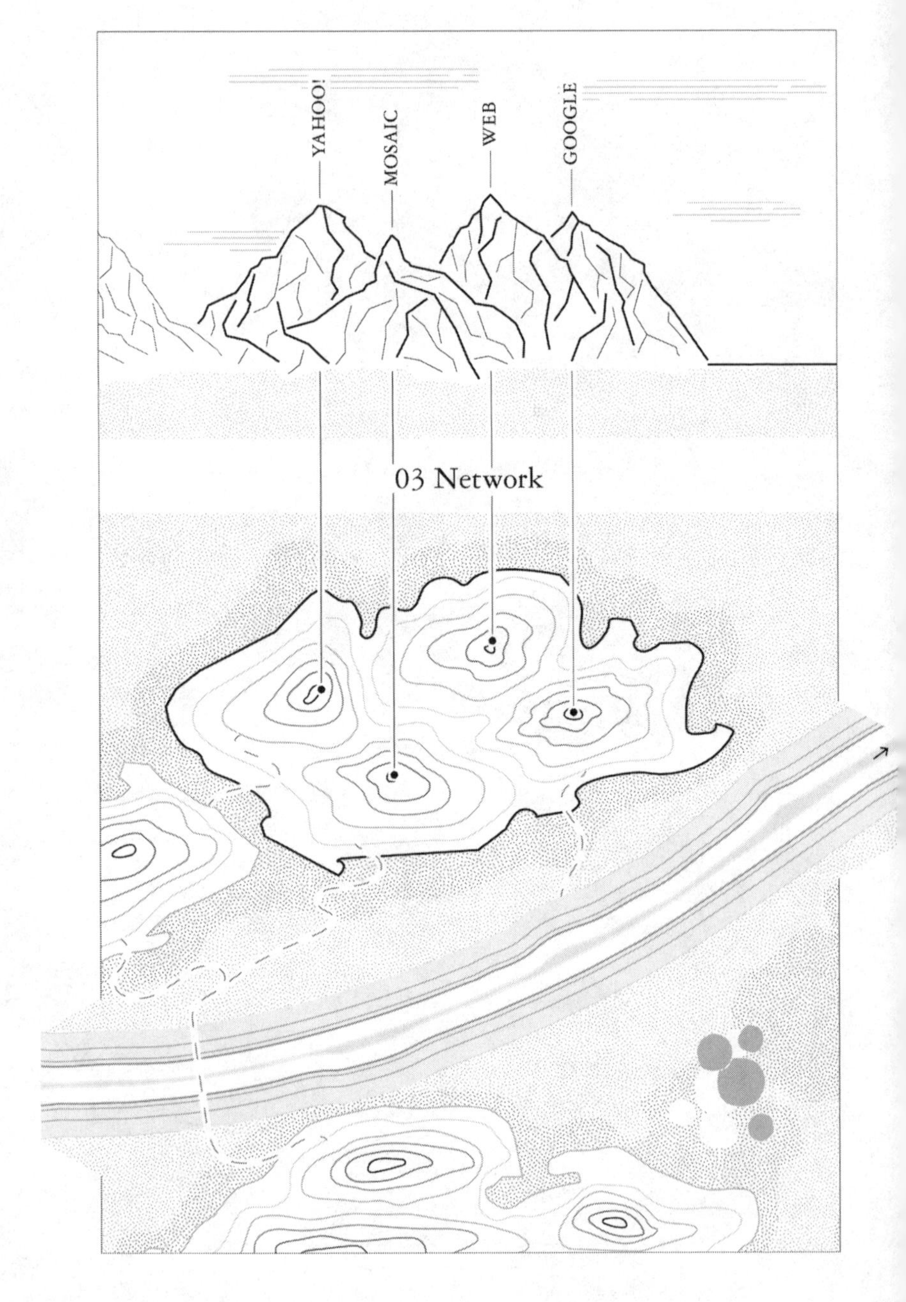
YAHOO!
MOSAIC
WEB
GOOGLE
03 Network

1995

• After photos, it was the turn of audiovisuals to be digitized. The first DVD went on sale. Once again, it was the Dutch company, Philips, against the Japanese (Sony, Toshiba, Panasonic). In two years, VHS technology was dead. May it rest in peace.

• Bill Gates launched Windows 95, the operative system that made all computers as user-friendly as Apple devices, but much more cheaply. There were no longer any alibis for putting off buying a home computer. If you didn't have one, you were really putting your head in the sand…

• eBay was founded, in California (again). A world marketplace where anyone could buy or sell anything. Their first-ever sale was a broken laser pointer.

1998

• Grand finale. Two twenty-four-year-old students at Stanford —Sergey Brin and Larry Page—launched a search engine they christened Google, another idiotic name. Now, it's the most visited website in the world. When they envisioned it, there were only 600,000 websites. They devised a way to find all the sites that contained, say, a recipe for *lasagna*, and to list them in order of importance in less than a second (*lasagna* was just an example; it worked if you looked up *hip replacement*, too).

The most astounding thing is that Google is still able to do this today, even though there are more than 1,200,000,000 websites in the world. If we wanted to adopt a metaphor from the sixteenth century, we could say that browsers like Google procured the frigates that helped you chart the high seas of the internet, portals like Yahoo! suggested the routes you should take and flagged the dangers, while Brin and Page found the way to calculate latitude and longitude and gave every single navigator a globe listing the world's ports in order of importance, convenience, and volume of trade. They told you where the best meals could be found, where the price of pepper was lowest, and where the finest brothels were. It will not surprise you to know that, today, Google is the most influential brand in the world (whatever that means).

Aside from the massive economic consequences of these inventions, some mental shifts also took place. This is important evidence for our profile of the newly born civilization. Unprecedented logic and unheard-of mental postures: an absolute novelty. It will be interesting to discuss this in the "Commentaries" that follow this chapter. For now, let's stop here and look at what's clear for all to see.

FINAL SCREENSHOT

Can you see the backbone of the mountain range? This is the Classical Era of the digital revolution. *Space Invaders* was the first little hill, more symbolic than real. These are real peaks. They're spectacular, right? Shall we try to summarize the story so simply that even a child would understand? Yes, let's.

Here we go.

The digital revolution came about as a result of three long moves that created a new playing field.

1. Digitize text, sounds, and images: reduce the substance of the world to a liquid form.
   This move went from CDs to DVDs, passing by MP3s, over the period from 1982 to 1995. More or less the same period as the development of the PC.

2. Construct personal computers.
   This move started many years before and only became visible in the early eighties—with the three PCs cited. It became irreversible in the mid-nineties with the arrival of Windows 95.

3. Allow all computers to communicate by placing them in a network.
   This move started with ARPANET in 1969 and, owing to the invention of the web, it finally achieved its aim in 1998 with the invention of Google.

Let's summarize even further. What happened in the Classical Era was that data containing the whole world was rendered liquid (A). Then, an infinitely long tube was built so that this liquid could be funneled at vertiginous speed into everyone's house (B). Finally, elegant faucets and washbasins were invented as terminals for this immense aqueduct (C).

By 1998, these tasks had been completed. There was room for improvement, of course, but it was all done. What we can say with relative safety is that human beings in the developed world, sitting in front of their PCs on any day in 1998, had at their fingertips a faucet that was easy to use and that gave them access to an immense aqueduct. More importantly, not only could they get as much water as they wanted, they could also inject water into the system. Or fizzy lemonade. Or whisky, for that matter. It was quite incredible and absolutely unprecedented. Our task, which is not only significant but also fun, is to see what those human beings did when they first got their hands on that faucet.

Basically, they used the immense aqueduct to circulate three different things: personal information (emails, research); merchandise (Amazon, eBay, video games); and aqueduct maps (Yahoo!,

Google). Naturally, if we went back to those years and looked in greater detail, we would find an infinite number of uses for the internet. However, what we are doing now is establishing which geological formations making up the backbone of the revolution sprung up when, and which of these formations grew into real mountain peaks. What we see is very clear: documents, merchandize, and maps.

The earliest explorers, navigating between continents in the sixteenth century, were circulating pretty much the same things. In other words, it was a very traditional strategy. A classic opening move, one would say, in the game of chess. Even in its most hidden move, which turned out to be the most influential. There was another thing that sixteenth-century merchants circulated around the world: *God*. Missionaries. A certain *way of life*. A certain way of being in the world. The digital revolution does precisely the same: it starts by establishing a certain way of being in the world, creating figures of thought and previously unchartered logical moves. A different idea of order, of how to grasp reality. Not exactly a religion, but something very close: a CIVILIZATION.

We can recognize it if we observe and study those early moves—archeological remains—from close up. In those moves there is something that recurs constantly, with common somatic features, tics repeated over time. Evidence of a mutation. Strange human footprints, which have never been seen before.

If you want to know more, read the "Commentaries" that follow. They are fascinating. They break the rhythm of this reconstruction of the backbone of the digital revolution a little, but they will also help you *truly* understand it.

Or, now that I think about it, you could always throw the book into the wood stove. I'd totally understand if you did, but I will now move on, without caring one way or another, to the "Commentaries." I love that word. It's so vintage.

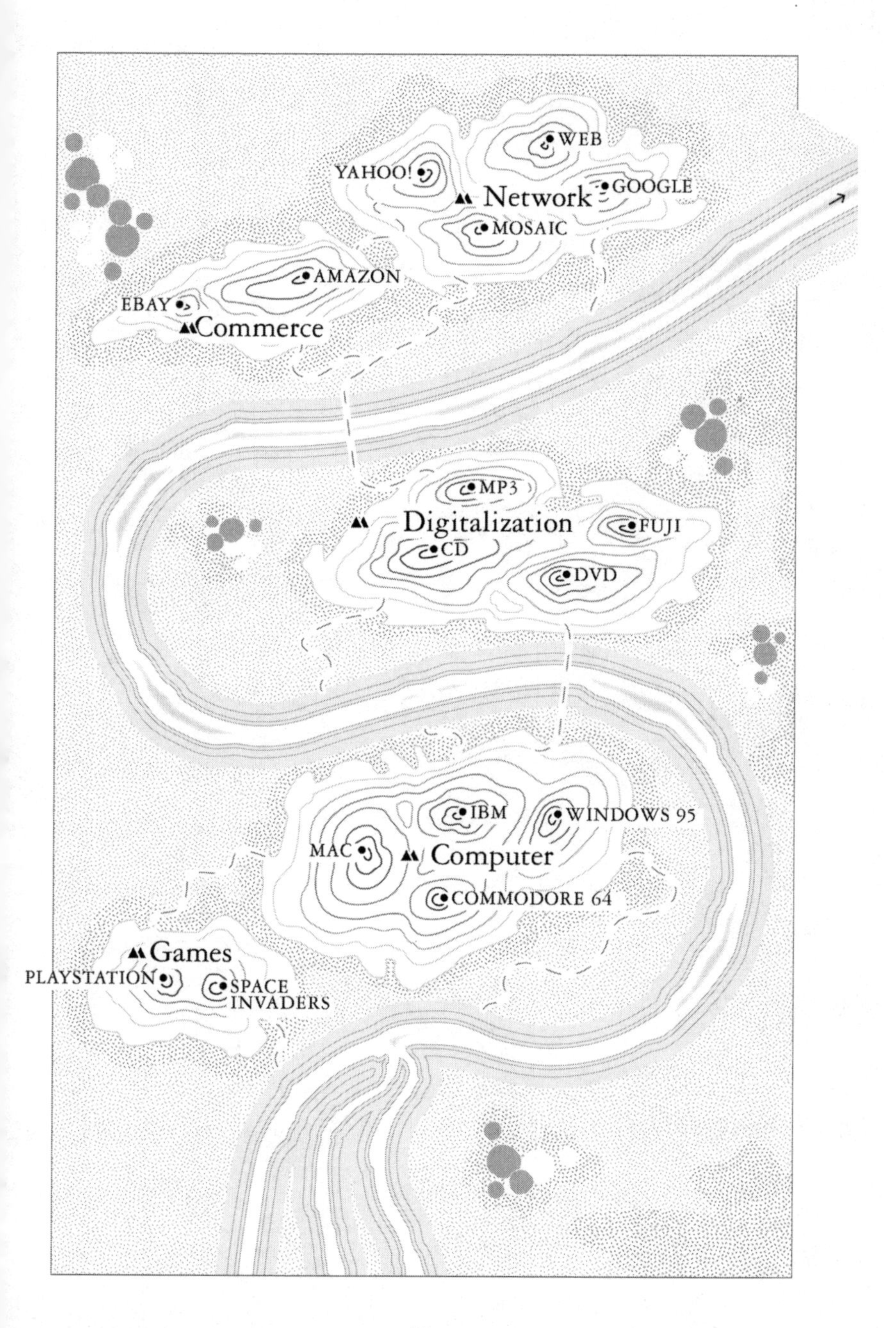
WEB
YAHOO!
Network
GOOGLE
MOSAIC
AMAZON
EBAY
Commerce
MP3
Digitalization
FUJI
CD
DVD
IBM
WINDOWS 95
MAC
Computer
COMMODORE 64
Games
PLAYSTATION
SPACE
INVADERS

COMMENTARIES ON THE CLASSICAL ERA

THE DEMISE OF MEDIATIONS

Reconstructing the backbone of the digital revolution is useful because, once you see the whole range of mountains, you can go and dig. Like a geologist or an archeologist. Dig for what, you may ask?

For fossils. For footprints left by those human beings. Clues.

The first clue is visible to the naked eye; even a child would see it. THEY WERE ALL ORGANISMS THAT SKIPPED STEPS AND SOUGHT DIRECT CONTACT WITH THINGS. That was the way they moved, by skipping steps.

DIRECT CONTACT

Merchandize is the easiest example: by selling books online, Amazon skipped a step (well, more than one, but let's stick to what is visible to the naked eye) and put bookshops out of business. The concept launched by eBay went further and skipped all the steps, eliminating the need for stores altogether. By all rights, it should have put Amazon out of business too, because the merchandize went straight from person to person, cutting out the mediators. Dealers no longer had a role.

The path emails took was similarly linear: straight from sender to receiver. Where did all the mail carriers, stamps, and the legendary postal system end up? They seemed to have vanished. Envelopes, letter paper? Poof! They were gone. Google was the most glaring example. It did away with all the intermediate steps. There was no longer a caste of wise men who knew where to procure knowledge. An algorithm, working invisibly, knew. It took you straight to whatever you were looking for.

This was a new phenomenon. Skipping steps was not part of the game strategy in the era of classic opening moves. It is therefore worthwhile examining it closer up.

TIDES

It's interesting to see what stays on the game table when you get rid of all possible mediations. At first glance, what remains is a kind of distant, almost mythological TABLE MANAGER who enables players to use the available options, oversees the serviceability of the system, and directs traffic, while creaming off a percentage of the profit. Basically, like eBay, or Yahoo! This manager is completely impersonal, or maybe I should say supra-personal (not an individual but, rather, a nebula of individuals). Often, it's an algorithm, as in the case of Google, or a computer protocol, as in email technology. It's little more than a system of rules, an organized space, a clear table, an open playing field vaguely controlled by some kind of remote entity. The first impression is undeniably of complete freedom.

However, if you look more closely, it becomes clear that in the apparent emptiness of this open playing field, a new force can be recognized. This force is so new that it is hard to zoom in on it clearly: it is like an almost imperceptible current. Tidal flows, generated by the movement of users circulating in the empty space. Amazon customers or people using the Google search engine. They move around, navigate, and leave traces behind. Right from the start, managers of these open spaces realized these traces were important and took great care not to erase them. In fact, they soon learned how to record them, organize them, make them legible, and, eventually, use them, giving them value. The most blatant example is again Google. Their aim was to compile an index of ranked web pages that would provide the answers to your questions; it was perfectly logical, of course. The hard part was choosing which pages should be ranked in first position; ultimately, the ones designated as the best. Traditional logic would have dictated that the company hired experts to indicate, question by question, which sites best fulfilled the criteria. But Google was working with such a high volume of users that it would have been impossible. In any case, it had already skipped traditional logic. Its instinct was to skip all the intermediate phases, cut out the mediators, and aim for direct contact with the world. So, no experts. Then what? Well, they went on to make one of those revolutionary moves that is at the core of the mutation we are studying. They decided that their clients' choices should dictate what should be at the top of the list and what should be at the bottom. By capturing the traces of every single user's journey through the network, they were able to identify which currents were stronger and which were weaker. With very little correction, this was to become the geography of knowledge. THE PLACE WHERE MOST PEOPLE WENT WAS THE BEST PLACE. The result is that, today, Google is consulted as if it were an oracle even though it is not an expert in anything. This is because it can give you a precise, overall picture of the views of millions of people. The principle was a game changer: the opinion of millions of incompetent people, if you know how to interpret it, is now considered more trustworthy than the evaluation of an expert.

We can observe, then, that when booksellers, mail carriers, dealers, experts—in sum, all the high priests—have been disposed of, what remains is the vigilant presence of a remote system and, sometimes, the currents produced by huge collective waves. The result is a kind of TIDAL EFFECT. Single individuals may swim freely in a safe, organized seascape, with no experts to tell them what to do. At the same time, however, there are currents created by immense collective tides that sweep them along without their hardly noticing. A fly buzzing around happily in the carriage of a moving train would be pretty much in the same situation. Would we be able to honestly say the fly was *free*? I don't know about the fly, but if we go back to the human beings in this new civilization, with their tendency to skip steps and seek direct contact with the world, I think we could safely say that they enjoy a certain degree of freedom. At least as much as they enjoyed before the digital revolution. In those analogical days, the tides were caused by the flows of mass ideologies that were impossible to escape (church, party, etc.). In the Classical Era of the digital revolution, by contrast, they were triggered by the mass movements regularly recorded by the dominant players of the revolution. It's hard to say which is preferable. There will be a time when we will be able to decide, and it will be interesting. For now, however, I'm just highlighting an incredibly powerful effect of the desire for direct contact: the twilight of high priests.

DEMOLITION OF THE ELITE

If you cut out mediation, the caste of mediators is inevitably disqualified from the game. In the long run, you kill off the old elites. The mail carrier, the bookseller, the college professor: they were all high priests, each in their own way. They were members of an elite who were recognized as possessing a particular skill or authority, and, ultimately, wielding that power. By constructing a system that disqualifies the elite from the game, replacing them with protected environments where human beings are placed in direct contact with things and are persuaded to float around in pursuit of the tides created by an unfathomable mass intelligence, something momentous happens. There is, apparently, a new world where the elite no longer counts, a planet in direct drive, where intentions and collective intelligence can be expressed without intermediary authorities. The inevitable consequence of this is that vast swaths of the population are convinced that mediators, experts, or high priests are no longer necessary. Many of these people believe they have been hoodwinked for centuries. They look around and, animated by a comprehensible vein of resentment, seek out the next mediator to destroy, the next step to skip, the next priestly caste to disqualify. Once you've discovered you can happily live without your old travel agent, why cling to the idea of a family doctor? In the area of politics, which is hugely overexposed to these mechanisms, the current inclination of voters to go for some form of populist leaderism that tends to do away with the mediation of traditional parties and even of traditional logical arguments, gives a startlingly clear idea of the phenomenon. Politics is just an example, of course, and it is not even the most important one.

SUMMARY

Since its foundational Classical Era, the backbone of the digital revolution has borne witness to the existence of organisms with a strong instinct for settling in a world that is activated by direct contact, skipping every possible step, and reducing to a bare minimum the role of mediation between human beings and things, or between one person and another. Individuals—who are relatively free to move around, but almost entirely lack reference points—end up using the millions of traces left by other individuals to guide them, because they can be read, organized, and translated into certain facts. Every single human being who thinks they are swimming freely eventually floats on the same mass tide. By the end of this process, these human beings get to experience life without any high priests, experts, or fathers. They love it. THEY COME OUT WITH AN INVIGORATED SENSE OF SELF.

DEMATERIALIZATION

Let's go back to *Space Invaders* for a moment. Do you recall the table soccer-pinball-video game sequence? If you remember, we outlined that gradual slide toward a reality with no attrition, that slow melting away of the body, that progressive slippage of the tangible to the immaterial. I think we can now safely say that the same sensation can be found quite regularly in all the ridges of the backbone corresponding to the Classical Era.

Digitalization melted down data until it was so light it was immaterial. Texts, sounds, and images were boiled down to nothing so that they could be conjured up from nowhere with the aid of devices that got smaller and smaller, almost as if they wanted to withdraw from reality by occupying less and less physical space. In the meantime, computers dematerialized practically the whole world, reducing everything to a screen that could be accessed by touching keys and moving a mouse (which, in its turn, became too material and subsequently vanished). By the same token, writing and mailing a letter had become something you could do sitting at your desk with a key command. Buying a book on Amazon or a used bike on eBay was turned into a process that only became real, tangible, or material when the package was actually delivered. Before delivery, the whole process was dematerialized to the point that the whole thing could be a fairy tale, entailing representations of objects that bore no relationship to the objects themselves, except as a well-intentioned imago. Not to mention the

PlayStation, which achieved the visionary dream of *Space Invaders* by converting the act of driving a racing car (or shooting at an old lady or shooting a penalty) into a lifelike experience as long as it was not real at all. Finally, of course, there was the web itself, and before that, the internet, which was—and still is—an entity perceived as IMMATERIAL even though it is "real." It is not real in the way railways or even shipping routes are. Does it weigh anything? Does it occupy space? Is it in a place? Can it break? Does it have edges? These questions have no answer, but neither could we say what space invaders were made of, could we? Did anyone know? No, nobody knew.

Dematerialization.

I'll try to translate (when I say translate, I don't mean that I'm about to express the same concept in a language that's easy enough for you, poor idiot, to understand. What I mean is that I'll try to translate to myself; I'll try to convert a collection of data into a legible form with the roundness of complete meaning). As I was saying, I'll try to translate. From the Classical Era of the digital revolution onward, increasingly vast areas of the real world were becoming accessible by means of immaterial processes. Let's say it was becoming an experience whose material elements had been stripped to an absolute minimum. It was as if the instinct of those early organisms was to reduce to the greatest extent possible their contact with physical reality in order to make their relationship with the world and with others more fluid, purer, and more pleasurable. Or again, as if they had decided to harvest wholesale quantities of reality and store them in granaries where they reduced their weight, made them easier to consume, and preserved their nutritional value for the winter or in case of a siege. Still again, as if they were attempting each time to isolate experience and translate it into an artificial language in order to keep it safe from the variables of material reality. Or, finally, as if they felt the urgent need to melt down all their material wealth into gold bars that were easy to hide and transport, malleable enough to adapt to any hiding place, and so shatterproof that they would survive any explosion.

A spontaneous question comes to mind: what were they scared of? Who were they *escaping* from? Were they preparing to establish a *nomadic* civilization? And if so, why?

AUGMENTED HUMANITY

Given this tendency to dematerialize experience and melt the world down into lighter, more transportable forms, the web is the highest, clearest, and most visionary embodiment of the trend. We need to look closer to understand why.

A good start would be the very first web page in history: the page where Professor Berners-Lee explained what the web was. It is a wonderful archeological artefact.

Go to: http://info.cern.ch/hypertext/WWW/TheProject.html

The definition of the web (not for us, but for a world that had *absolutely no idea what it was*) was made up of TWENTY-ONE WORDS. The whole first page contains less than two hundred words (lightness, concision: think of the transition from table soccer to *Space Invaders*). To make up for this brevity, the sixth word ("hypermedia") was printed in blue and underlined. By clicking on this word, you ended up on another page, which was also very brief. The first line offers in ten words the definition of hypertext: "Hypertext is text which is not constrained to be linear." Great. Text that is free of the chains of linearity. A text that could be in the form of a web, a tree, a leaf, whatever you want to call it. A text that explodes into space and doesn't need to be written from left to right or from top to bottom. As you grasp what it means, you are already inside it; you are moving like hypertext. You keep on clicking on those blue words, and they lead you gently but quickly along a diagonal path that almost switches back on itself, following a movement you've never made before. Wandering aimlessly in this way, experiencing a lightness of being you've never felt before, you bump into words or phrases that describe what you are feeling. One of these is: "There's no 'top' to the web. You can look at it from many different points of view." For a civilization that had been accustomed for centuries to ranking everything from top to bottom and that had always tackled problems by organizing them in order of size, this man was saying that the web was a world with neither a beginning nor an end, neither a before nor an after, neither a top nor a bottom. You could enter from anywhere, and "anywhere" would always be the main gate, but never the *only* main gate. You can see the reflection of a great mental revolution, can't you? It wasn't merely a technical question, an issue of ordering material; it was a matter of mental structure, of the way thoughts moved, and the way the brain was engaged. Another phrase sounds highly significant to my ears, considering its splendid sobriety: "Hypertext and hypermedia are concepts, not products." Professor Berners-Lee knew who his audience was. He knew he had to express things clearly, in explicit terms: he was describing a *way of thinking, not a tool that you buy and use while continuing to think in the same way as before*. It is a way to move the mind, but it is up to you to choose *your own way* of moving your mind.

For people choosing their own way, the web gave them a sensation that we must register here as a fundamental element of the transformation, because it marked the most evident difference between the internet and the web. In many ways, the internet, while it may have felt like science fiction, actually relied on a fairly traditional pattern of experience. I am here, I load a piece of information or merchandize onto some means of transport or other, and that item reaches another human being on the other side of the world in a second. Nice, but really no different from a telegraph, with all its limits. It didn't offer a mental experience that was fundamentally different.

Things change drastically when you start surfing the web. Whatever really happens in the technological belly of the web, the impression you get when you are surfing is that YOU are moving, not the things around you. You are the one who can be catapulted to the other side of the world in an instant; you can look around, grab whatever you want, fly around all over the place, seize something else, and be home in time for dinner. We say we *send* an email on the internet (I stay in one place and the email is the one to travel), while we say we *surf* the web (I'm the one who's moving, not the waves).

The difference is highly significant in terms of mental models and self-perception. The entire digital revolution, as we have found out, was obsessed with melting the world down into light, fast, and movable fragments. Yet, it is easy to see how the web raised the stakes enormously. It did not limit itself to dematerializing *things*; it dematerialized *human beings*! Technically, it transported packages of digital data, but at a level of sensation, of impression, what it did was make *us* as light, fast, and movable as those packages of data. When the computer was switched off, we went back to being the same old elephants, but when we were surfing the web, we became animals with the same design as our digital products, with the same hunting techniques.

This has a consequence that may sound a little sinister but that I propose we consider with the care it requires. Clicking one hypertext after another, people surfing the web began to develop a perception of themselves as HYPERHUMANS. I wouldn't like for you to interpret this in any way that may be reminiscent of the Nazis or even of Marvel Comics. It's not that you felt as though you were a god or a superhero with extraordinary powers. It's that you felt like a hyperhuman: that is, A HUMAN BEING WHO WAS

NOT OBLIGED TO BE LINEAR. Someone who didn't have to stick to one mental space; who refused to let the world dictate the way their thoughts should be structured, or the way their mind should move; who didn't always have to use the main gate.

A new human, one would be tempted to say. This is where, exactly at this point in the discussion, the digital revolution hints at the fact that it was born of the mental revolution. This is the first time that a hypothesis has been put forward that A DIFFERENT HUMAN BEING was behind the development of digital technology, and that A DIFFERENT HUMAN BEING would most likely be the result.

This point is of vital importance.

Let's make a special effort to consider it a perfectly innocent one, because that is what it was. It was a perception of a kind of augmented humanity. Forget Twitter, Facebook, and WhatsApp; forget even artificial intelligence for now. We'll get there later, but for now, set them aside. None of these applications existed at the time. At the time, there was the perception of being augmented humans with the freedom to reject cumbersome, inflexible, sluggish movements. You need to feel on your own skin what that unexpected melting of the world must have felt like, how it suddenly lost all its abrasiveness. The yearning of the space invaders. Except that, in this case, it was no longer a game; it was life.

What kind of death were people escaping from when they decided to live their life in a way that they'd never seen or experienced before?

THE OTHERWORLD

The web did not only allude to NEW HUMANS, it also opened up their natural habitat. This is the crucial issue.

How can we say what the web did in words simple enough for a child to grasp? IT CREATED A DIGITAL COPY OF THE WORLD. The web was not created in an elite laboratory; it came about as a sum of an infinite number of little acts. It was a kind of otherworld that gushed out of every single user's craftsmanship. While it may have seemed a little artificial, it was INFINITELY MORE ACCESSIBLE. Membership requirements were trifling: you had to be able to afford a computer but, apart from that, there were no other cultural or economic hurdles in your way. You could move around freely in the otherworld at no cost whatsoever. This was almost inconceivable.

Moreover, the copy of the world created by the web offered a type of reality that was much smarter than the reality you had to deal with in the everyday world. You could travel in all directions, move around in complete freedom, organize your material experience according to an infinite range of criteria, and do all these things in no time at all. In comparison, true reality—the first world—was cumbersome, sluggish, abrasive, and controlled by a principle of obtuseness. Again, the difference between table soccer and a video game.

It may feel risky, but I think we need to go one phase further and admit that the mental model that the otherworld was developing promised something that was better suited to our abilities. I may go as far as saying something more NATURAL. At a closer look, the system of hyperlinks simply replicated the brilliant workings of something we know very well: OUR BRAIN, which is often forced to proceed in a linear fashion but was probably not born that way. Left to its own devices, our brain continuously opens links; it keeps many windows open simultaneously and never gets to the bottom of anything, because it tends to make leaps sideways from subject to subject. All the while, it preserves, in a kind of hard disk, the memory and the map of the journey. Just think of the effort children have to make to focus on their homework, a math problem, or reading a page of a book. We all know that if they were not forced to be linear, their minds would be flitting about very much like the web suggests they should. In the old days, this nonlinear way of thinking was stigmatized, without appeal, as a mental technique that would never solve problems or trigger experience. Then, the web appeared as a witness for the defense, stating that nonlinear thinking was precisely what would help us solve a whole range of problems and would give us a remarkable and meaningful experience of the world. Actually, it didn't really state it. All you had to do was surf the web for a while and it demonstrated its point to you. This shift could not go unobserved, of course. The message was that the side of you that was instinctive, anarchic, or undisciplined was no less valuable in terms of your powers as an explorer than the pathetic navy officer they were trying to train you to be every day at school. With the proviso—and this is the real point—that you accepted that there were other oceans where reality had been duplicated and converted into a different format that was more suited to your mind. This was where you should have gone surfing. In the waters of the otherworld.

In these utopian traits—providing human beings with a playing field more suited to their instinctive capacities and more accessible to whoever wanted to play—the web was trying to bring to fulfilment impulses that had been around for a long time. We can see this in some processes that had nothing to do with the digital revolution but that attempted to achieve the same result as the web by changing the habits of the real world, not its copy. I'll cite four examples just to give you an idea. We owe them a great deal, as they represent a fantastic pre-web, pre-digital world. Here they are (the dates refer to Europe; the US is a world apart):

- Supermarkets in the fifties
- Television since the sixties
- Dutch Total Football in the seventies
- Low Cost Flights in the eighties.

It's not hard to see how the web borrowed something from all four of these phenomena. There was a mixture of accessibility, freedom, and speed in these models that represented a break with decades of clogged, sluggish, discriminatory systems. Rigid boundaries were melted down, and whole sectors of experience (shopping, pastimes, information gathering, playing soccer, traveling) were suddenly freed from all the impractical, detrimental drawstrings of the previous era. In these contexts too, a certain reduction in quality—even in reality—was accepted as par for the course. On Ryanair flights, you couldn't choose your seat; at the supermarket, nobody would ask you how your kid was doing in school; the Dutch national soccer team practically never won a match; and watching TV—compared to going to the theater, the opera, or even the movies—was clearly a fallback activity. And yet, the lure was irresistible. The attraction had something to do with a widening of horizons, the loosening of rules, the demolition of senseless mental blocks, and the re-vindication of a new egalitarian approach. The web unconsciously inherited that idealistic impulse, and brought it to triumphant completion, by means of a strategy that was brilliant and risky in equal parts. Rather than try to modify the world directly, it proceeded in stealth, with a surprise attack from behind. In an act that had unparalleled consequences, it invited everyone to duplicate it, representing it by means of a myriad of digital pages. Once a copy of the world was made, it became possible to fly Ryanair for ten cents; play any position on the soccer field, as Cruijff did; bring the whole world into our sitting room, as television did; and wander around the products of the world pushing a shopping cart. It was, of course, an irresistible move.

Checkmate.

Another way of putting it is that, by applying the logic of MP3s to the red-hot matter of experience, the web offered human beings a compressed version of the created world that was written in a more comprehensible language and refashioned in order to eliminate the walls that had previously made experience a luxury product for the few. In short, it irreversibly altered the way the world was formatted.

Please turn your cell phones off, leave your partner alone for a while, and give me a minute of your full attention. The operative phrase is this:

It thus irreversibly altered the way the world was formatted.

You need to understand this concept.

What function do millions of web pages currently inhabiting a virtual non-place right next to the real world perform? Together they make up a second heart, adjacent to the first heart, that pumps reality. This is one of the web's most brilliant ideas:

▸ It equipped the world with a second pump, so that the flow of reality would circulate in a blood system with two adjacent hearts, pumping in harmony, correcting each other, and standing in for each other beat by beat. ◂

Do not mistake me: I'm not saying that the habitat of digital hyperhumans is the otherworld of the web. It's far more sophisticated than that. Their habitat is a system of reality where a twin pump has melded into a single one because the distinction between the real and the virtual world has become blurred; this single pump, in its turn, generates its own reality. This is a more precise description of the playing field where the new humans operate. Their habitat is tailor-made for them, and civilization has crystallized itself around them. It's a system where the world and the otherworld are intertwined. Together, they produce experience in an infinite, permanent creative process.

This is the scenario today. Inaugurated in the early nineties, it has been passed down to us after undergoing a myriad of improvements that we will only discover over time. This is the game that awaits us every morning. If you don't learn the rules, you will face pitiful defeat.

WEBING

The web first launched a digital otherworld, and then made it interdependent with the first world. The result was a single system of reality driven by a double pump. Once we learned this trick from the web, we replicated it in different forms, many of which had little to do with the web. For example, you don't need the web to play *FIFA 2018* or send a WhatsApp message or read a book on Kindle. We are not on the web when we look for a dinner date (euphemism) on Tinder, nor when we listen to music on Spotify. And yet, all these acts are simple variants on the revolutionary theme invented by the web: that is, they all bounce between the real world and the virtual otherworld, weaving a pattern that we legitimately call REALITY. In this sense, the curious fact that we tend to think of everything as being on the web, and often do not even distinguish between the web and the internet, betrays a childlike view that reveals a great deal. Everything is on the web. Whatever we do is webing; whenever we produce reality by making the two worlds pump together, it is webing; when we go into the otherworld with our app in order to manage our everyday, material world, it is still webing. We are always webing. This is our way of life, our way of creating meaning and accumulating experience. From this point of view, we are truly unheard-of specimens of humanity; we already were in the earliest days of the digital revolution when, in the dawn light of the Classical Era, its foundations were laid.

In an attempt to tie things up with a neat, lasting definition that we can carry around in our pockets to ward off danger, it would be very useful to go back to a clue we first found in the innocent little game of *Space Invaders*. Do you remember the *human-keyboard-screen* posture? We saw it as nothing more than a way to be *physically* in the world, and yet it was quite revolutionary. Now we know that it is a symbol for a fairly complex act. That is, the act of creating a system whereby the world and the digital otherworld are able to communicate and establish—by means of that *human-keyboard-screen* posture—a new system of reality powered by a double pump. It looked like nothing more interesting than a way of being, but it was actually a brilliant way of conducting one's existence. Our existence. We are that human being. The *human-keyboard-screen* image is the logo of our civilization.

It has the same essence of an icon that for centuries represented another type of civilization: man-sword-horse. It was a warfaring civilization, and that posture summarized everything they had to say about life. Their playing field was the physical world; the sword and the horse were the tools they used to leave their mark on it. We are the *human-keyboard-screen* people. Our playing field is more complex because there are two hearts, two generators of reality: the world and the digital otherworld. The logo represents the precise moment when we are sitting in the former and traveling through the latter. Webing.

It is a very effective logo. Why don't you embroider it on the back of all your fears?

MACHINES

The *human-keyboard-screen* logo pins down an unequivocal truth we are loath to accept. That is, none of these developments would have taken place if people hadn't consented to experience a good part of their lives through machines.

Of course, human beings were not absolute novices in this regard. Galileo's telescopes were machines; using them to increase human knowledge felt like a great idea at the time (with the exception of a handful of bishops and popes, of course). More recently, people happily accepted communicating by means of a machine—the telephone—that eliminated at least half of their potential experience of the act (being face to face and looking at each other). Nevertheless, the only complaints people ever made was when the line was disturbed. In short, human beings previously acquired a certain experience of machines that modified their experience. However, the case of computers and the otherworld was different: all of a sudden, thanks to a machine, you could generate and inhabit an amplification of reality, a multiplication of the world. It wasn't exactly like warming up a glass of milk in the microwave. What happened, de facto, was that computers not only helped you manage reality, they also generated, at your command, a new reality that complemented the other. This was a serious departure from the past. Once we consented to go down this road—the road that led to machines being used to correct or recommence the Creation—there was probably no way back, which may be what frightens us today. It was a scary decision, especially when we made it with unabashed nonchalance, shooting at little Martians or buying neckties online. Yet, hardly anyone cares; at best, they see it as a distant, original sin that one day they may expiate. The reaction is irrational, but it may explain many of the senseless qualms and fears that accompany us (such as the idea that we'll be replaced by robots!).

The thing that strikes me in particular, continuing with my analysis, is that the very same human beings that sought direct contact whenever they could, and systematically skipped all the steps, were the ones who spawned an idea that went in the opposite direction: that is, using machines to augment their experience. It's odd, isn't it? It's a logical tangle that is hard to unpick. It almost certainly reveals something about those human beings, but what exactly? It reminds me of something I wrote a few pages back (yes, I'm a little self-referential, but so what?): "the PlayStation achieved the visionary dream of *Space Invaders* by converting the act of driving a racing car into a lifelike experience as long as it is not real at all." *A lifelike experience as long as it is not real at all*. Here we are with another logical tangle. Could it be related to the other one? Do they both express something I've been unable to unpick?

Very probably. As soon as I try to come to grips with the matter, I realize my mistake: I'm still adopting the old, pre-revolutionary mentality. It's not surprising; given that I was born in the middle of the twentieth century, how else am I supposed to think? I need to adopt the collective mentality of the human beings who gave rise to the mountain range I'm trying to analyze.

The two problems may look like logical tangles, but I must convince myself that they are not. Thinking of computers as mediators may be a reasonable thing to think for a man born in the mid-1900s, but it's nonsense for millennials who see their computers as extensions of themselves, rather than as something that mediates their relationship with the outside world. A smartphone is no different from a pair of shoes, a lifestyle choice, or music preferences; they are all extensions of their egos. The instinct to cut out every kind of mediation does not conflict with an obsessive reliance on machines for the simple reason that, for millennials, those machines DO NOT MEDIATE. They articulate their way of being in the world. Similarly, wasting time by attempting to distinguish the real world from an unreal world in the experience provided by a PlayStation is a dubious indulgence when the real world and the digital otherworld flow together in a single circulatory system and together produce a single system of reality. For millennials, trying to find the demarcation line between the real and the unreal in a game of *FIFA 2018* would be like trying to separate peas from carrots in a vegetable soup, or like trying to establish whether angels are male, female, or transgender. They're angels. Who cares? And vegetable soup is vegetable soup, for crying out loud. So, if I go back to that phrase I thought was so brilliant ("the PlayStation achieved the visionary dream of *Space Invaders* by converting the act of driving a racing car into a lifelike experience as long as it is not real at all"), I can now see that thirty years ago it might have deserved some praise but, objectively speaking, today it is clear that it is a beautifully written piece of bullshit.

I must admit, it's irritating to have to admit it.

I think I'll go and open a can of beer.

MOVEMENT

One more thing: the last, but still vitally important.

At the end of the day, when you zoom in and look more closely at all the moves that led to the Classical Era of the digital revolution, you'll see they have one chemical component in common. It can be found literally everywhere; it dominates all the others; and it somehow *preceded* all the others. This is, THE OBSESSION WITH MOVEMENT. These were people who dematerialized whatever they could get their hands on; they strove to make each and every created thing as light and mobile as possible; they spent their time constructing immense communications systems; they didn't stop until they had invented a cardiovascular system that allowed everything to circulate in every possible direction. These were people who experienced linearity as a constriction, who demolished any mediation that slowed things down, and who preferred speed over quality every time. These were people who went as far as building a digital otherworld in order to make it impossible for the world they were living in to lie around doing nothing, unchallenged.

Jeez! What was their problem?

They were on the run. That's the problem. They were escaping from a century that had seen the most horrific bloodbaths in the history of mankind, which affected everyone without exception. In its wake, there followed a remarkable series of disasters. When you zoom in and look more closely at that series of disasters, you'll see that they had one chemical component in common that can be found literally everywhere and that dominates all the others. That is, AN OBSESSION WITH BORDERS, AN IDOLATRY FOR ANY LINE OF DEMARCATION, AN INSTINCT FOR CREATING A WORLD ORDER WITH PROTECTED AREAS THAT DO NOT COMMUNICATE. They may be borders between nation states, ideologies, high and low cultures, or even a superior human race and an inferior one. Drawing a line and making it impassable was—for at least four generations—an obsession worth dying or killing for. The fact that the borders were artificial, invented, random, or senseless did absolutely zilch to stop the massacres. It's impossible to understand the digital revolution without reminding ourselves that the grandparents of the people who initiated the transformation had fought a war where millions of people died to defend the immovability of a border or to move a border a few miles—sometimes even a few yards—in one direction or another. A few years later, the blind isolation of the elite, the cultural inertia of nations, and the leaden torpor of news circulation meant the next generation lived in a world where it was possible to run concentration camps in Auschwitz without anyone knowing what was going on there, and to drop an atomic bomb without more than a handful of people being given the opportunity to debate the pros and cons of the operation. They grew up and went to school every morning in a world divided by the Iron Curtain and engaged in the Cold War, under the threat of a nuclear apocalypse. Everything was controlled in secret chambers by an isolated caste of the blinkered elite. We're not talking about barbarians in a pre-civilized world. These things took place in a corner of the world, the West, where an apparently advanced civilization had been handing down the art of cultivating lofty ideals and superior values for centuries. The tragedy was that the series of disasters did not appear to be the unexpected consequence of a vacant phase of that civilization. On the contrary, it was the lucid and inevitable result of its principles, of its rationality, of its way of being in the world. Anyone who witnessed the twentieth century knows it was no accident; it was the logical consequence of a certain thought system. Things could have gone better, but leaving that civilization to its own devices almost automatically led to that carnage. What could save you?

Setting everything into motion.

As soon as was humanly possible.

Boycott borders; pull down walls; create a single, open space where everything had to circulate. Demonize immobility. Consider movement a prime, necessary, totemic, undeniable value.

It was a brilliant intuition. The twentieth century taught us that if stable systems are left too long without change, they tend to degenerate into insatiable, destructive monoliths. An opinion was transformed into a fanatic belief; nationalist sentiment easily became blind aggression; the elite turned in on itself and behaved as a caste; truth became a mystic creed; falsity became myth; ignorance dissolved into barbarity; and culture was reduced to cynicism. The only remedy was not to allow all these different places in the world to become too immobile, to make sure they didn't become involuted or seek safety inside their shells. It was essential that people, ideas, and things were dragged out into the open and funneled into a dynamic system where there was hardly any attrition with the outside world and where the highest value, primary aim, and only foundation was facility of movement.

We stem from that decision.

Many traits of today's civilization can only be explained if you recognize MOVEMENT as its supreme—originally only—objective. It was the antidote to the poison that killed so many people so atrociously for centuries. There wasn't a great deal of discussion about its collateral effects or potential counter-indications. We were in a hurry, there was no room for doubt. There was the world to save.

Looking at the dates, it's easy to imagine how things must have gone. We were preparing ourselves for a while, and then we exploited the first window of opportunity history afforded us: 1989, the fall of the Berlin Wall. In those minutes, every wall came tumbling down, bringing the Iron Curtain down with it, and changing forever the meaning and value of walls, borders, and separations in Western minds. The window was wide open, and so we climbed right in. The digital revolution accompanied collective actions that were shifting in a similar direction: globalization and the foundation of the European Union were two evident examples. In a very short space of time, in fact, we broke an awful lot of chains and we imposed on ourselves a new game, on an open playing field, where movement was the most important skill. The antidote was already beginning to take effect. For example, an unheard-of situation for human beings in the developed world had begun to take root, inverting a trend that for millennia had been the mark of our civilization: lasting peace was for the first time seen as the best possible scenario for making money. Before, it had always been war.

At one point, suddenly, political instability or the risk of military intervention was seen as a curse, because it stems the

> flow of the planet, blocking the circulation of money, merchandize, ideas, or people. People want peace not because they believe in it, or because they are good people, but because it is in their interest—which is perhaps the only kind of pacifism that can survive any emergency. It has survived even though we have been desperately trying to split the world in two again with a border—a mythical border no less, with a certain history and fame—between the Western world and Islam. The clarity of purpose with which the Western world, and the powers that be, have reigned in their recourse to arms and kept in check the centuries-old rivalries of vast swaths of the population reveals a great deal about how deeply the antidote has penetrated the blood system. The art of movement seems to have reduced the risks to a minimum. Nowadays, we can allow ourselves the luxury of being choosy: questioning the impact of low-cost flights on quality tourist destinations or worrying whether Google has killed the new generation's ability to do geography homework. Swanks. Many of us may well be starting to wonder whether perhaps a few new walls might be built after all. There's a growing nostalgia for the elites. Short memories. There's a lot of work to do, and we still haven't finished.

Today, it's a matter of going back to the roots and understanding the first move, the one that precedes and explains all the others: WE GAVE MOVEMENT PRECEDENCE OVER EVERYTHING ELSE. It's important to understand this literally. If you make movement obligatory for everything that exists in this world, then movement will come back to haunt you at every layer of your experience, from the simplest to the most complex. There's no point in expecting millennials to do one thing at a time, aim for a permanent job, or stick to the same truth from one day to the next. Everything that requires stagnant immobility in order to have meaning stinks of the twentieth century and sounds vaguely sinister. This is why we prefer systems that set things in motion and don't allow them to sit there rotting. We have reached the point where we give value to things on the basis of their ability to generate or host movement. There is no truth or wonder that maintains its value in our eyes if it is incapable of circulating in the flow of a meaningful collective current. Thus, in order for things that happen to really exist, they need to take place along a trajectory. They very rarely happen at a stable point. Increasingly, there is no beginning and no end, and the meaning of an event is written in the changeable trail it leaves in its wake. Shooting stars. This is the way we move, without interruption, making us look a little neurotic and dispersive and, every now and again, making us doubt ourselves. We often consider this an effect of the machines but, once again, we must turn things the other way around. The truth is WE are the ones who chose movement as our primary aim. Machines are simply the tools that were tailor-made to pursue that aim. WE are the ones who wanted to move around the world unburdened. This is what we wanted when we started the revolution. There was a house in flames that we had to abandon in a hurry. We had an escape plan and a system for saving ourselves in mind. There were a few people who could see, in the distance, some kind of promised land.

MAPPA MUNDI 1

Digging mountains does not mean ascending them. Seeing them as archeological remains does not mean painting them in the sunset. We dig and work hard to find evidence of seismic movements that took place in remote times. We are looking for the beginning of everything. The work is clumsy and requires patience and the willingness to wait things out. Head down. WE have succeeded in doing this, and we now have under our eyes a first map of the earthquake that generated us. The first mappa mundi we were searching for.

What we can see is the dawn of a civilization and its reasons for existing.

They emerged out of a disaster. Two generations before them had lived, killed, and died in the name of principles and values that later revealed themselves as both sophisticated and lethal. They had done so under the unquestioned guidance of implacable elites, who had been carefully trained with lucid planning. The result was a century of unmitigated horror, and the first human community ever to invent a weapon that had the capacity to destroy them utterly. This was the paradoxical legacy that a civilization claiming to be super-refined left to future generations: the privilege to enjoy a tragic end.

It was in that moment that a kind of instinctive inertia urged some of those humans to escape. It was a mass exodus conducted in an almost clandestine fashion. They were escaping from themselves, their traditions, history, and civilization. They were chased by two enemies: 1) a troubling system of principles and values, and 2) an untouchable elite who presided over it. Both enemies were rooted in solid institutions that hadn't changed for centuries whose strength lay in their well-honed intelligence. They could be challenged in a face-to-face battle, which required them to produce ideas, principles, and values, very much as the Enlightenment philosophers in another period in a similar situation had done. An ideological battle in the field of ideas. However, the humans who had envisaged an escape plan had seen how often "ideas" had led to disasters and were therefore justifiably and instinctively suspicious. Moreover, they themselves were members of a male, technical, rational, pragmatic elite. Whatever talent they may have had was in the direction of problem solving, not of creating conceptual systems.

So, they instinctively tackled the problem by going back to the basics, INTERVENING IN THE WAYS THINGS WORKED. They started solving problems (any problem, even the problem of how to send a letter) BY SYSTEMATICALLY CHOOSING THE SOLUTION THAT PULLED THE CARPET OUT FROM UNDER THE FEET OF THE CIVILIZATION THEY WERE ESCAPING FROM. They didn't seek the best, or the most effective, solution; they chose the one that demolished the cornerstones of the civilization they wanted to be free of. They came from a civilization built on the myth of stability, permanence, borders, separations; they started tackling problems by adopting solutions that systematically ensured the greatest amount of movement, mobility, melding differences, and dismantling barriers. It was a civilization, teetering on the edge against a stable elite of high priests, entrusted with a reassuring system of mediation; they started adopting solutions that systematically skipped as many steps as possible, rendered mediation useless, and disqualified all the high priests from their role. They were fierce, rapacious, incredibly quick, and armed with a certain dose of urgency, scorn, and desire for revenge. It was more of an insurrection than a revolution. They stole almost all the available technology (they managed to steal the internet from the army, their enemy...). They used universities as warehouses, places where they could spend as much time as they needed to strip the institutions of everything potentially useful. They had no pity for the victims they left in their wake (nobody has ever seen Bezos weep for the bookshops he drove to bankruptcy). They had no ideological manifesto, no explicit philosophical views, no particularly clear guiding ideas. They weren't building a THEORY OF THE WORLD; they were establishing a PRACTICE OF THE WORLD. If you are looking for the founding principles of their philosophy, here they are: the Google algorithm, Berners-Lee's first web page, the iPhone screen. Things, not ideas. Mechanisms. Objects. Solutions. *Tools*. They were escaping from a destructive civilization and they did so by implementing a strategy that did not need special theories. The strategy consisted of solving problems by systematically choosing the solution that boycotted the enemy by privileging movement and dismantling every kind of mediation. It was devious, but relentless and hard to resist. Applied to every meander of experience—from buying a book to the way you take your holiday snaps to how you look up the meaning of "quantum physics"—it recklessly eroded the great palaces of power (schools, parliaments, churches) and ended up invading the world from below, liberating it almost invisibly. It was as if they had burrowed tunnels under the skin of twentieth-century society… it was inevitable that, sooner or later, there would be a landslide.

What we are now able to understand is that the serial application of solutions, chosen systematically because they facilitated movement and dismantled mediation, first generated new tools, which laid the foundations for digital etiquette: the digitalization of data, personal computers, the internet, and the web. We also know that using these tools later created completely unexpected and unprecedented scenarios which, in their turn, paved the way for a true mental revolution: the dematerialization of experience, the creation of an otherworld, access to augmented humanity, a system of reality with a double pumping system, the *human-keyboard-screen* posture.

Now, the question is: did they *want* this scenario? Was this the world they planned to create? Did they have an idea of what kind of human being they had created all this for? We can serenely answer all these questions in the negative. They didn't have an idea of what kind of world they wanted to find; all they knew was what kind of world they were escaping from. They didn't have a plan for humans; all they had was an urgent need to demolish the humans who had cheated them. Nevertheless, they possessed, in their problem-solving DNA, a formidable ability to keep up to date. Solution after solution, they found they were grappling with scenarios they had neither envisaged nor sought. They possessed a formidable ability to redirect them efficiently toward pursuing the ultimate aim of the insurrection. That is, to disarm twentieth-century mankind. In this respect, we must admit, they were brilliant. They made mistakes every now and again, of course, going down blind alleys or taking roads leading nowhere. Yet, in most cases (the backbone), the constant realignment with the aim of the insurrection is striking. They were pioneers, let us not forget it. And yet, they succeeded in redesigning a game board that was by no means a random choice; it represented precisely the game they had started playing. When they first started out, there was no way Google could have imagined it; however, as soon as they saw it with their own eyes, they realized that it was the product of their own mental revolution. They did not take long to adopt it as their strategic fortress, cutting off most of the enemy army forever. Take the story of the otherworld. It didn't take much, after generating it, to reduce it to a kind of warehouse for storing piles of more or less useless things. And yet, the founding fathers of the digital insurrection understood that, if taken seriously, the otherworld offered immense opportunities for declaring victory. If they managed to circulate reality there too, adding a digital heartbeat to the heart of the world, it would be much more difficult to limit human experience to that semi-paralysis that had seemed so indispensable for the disaster of the twentieth century. Similarly, the idea of augmented humanity, accessible to nearly everybody, undermined the very concept of elite from within. In some ways, it promised all the participants of the insurrection a share of the power that had previously been concentrated in the hands of the few. After all, the best way to get rid of a high priest is to allow everyone to perform miracles. In the meantime, the digitalization that had been forced onto all available information made the world so light that it guaranteed natural instability. It was a format born to facilitate movement, and it was possible to bet on the fact that it would go on to generate a continuous migration of all material in all directions. Try to draw an arbitrary border, separate human races, hide a nuclear bomb, or tell everyone Auschwitz was a labor camp nowadays. Good luck to you.

In short, maybe they didn't know where they were going, but once they were on their way, the path was clear. The people who designed the first personal computers would never have imagined the web, and those who came up with the MP3 idea wouldn't have been able to predict that their technology would one day be used on a platform like Spotify. However, a kind of collective compass lined everything up and decreed the escape a success. This helps us come up with one of the answers we were looking for (about time!). Do you remember? It was linked to our fears.

> *Are we sure that this isn't the kind of technological revolution that blindly imposes an anthropological metamorphosis that cannot be controlled? We have chosen our tools, and we like them, but did anyone—before the fact—bother to calculate the consequences of using them on the way we lead our lives, perhaps on our intelligence, and, in the extreme, on our very idea of good and evil? Do Gates, Bezos, Zuckerberg, Brin, Page* et al. *have a project for humanity in mind, or have they simply produced brilliant business ideas that have involuntarily, and rather randomly, produced a new kind of human being?*

Right. Now we can hazard a guess. No, the founding fathers of the digital insurrection did not have a precise plan for humanity, but they did have an instinctive sense of how to escape the disaster, and whatever they constructed followed this strategy. This means that the foundations of the civilization they launched were firmly based on its original motivation. This gave it the coherence and harmony everyone recognizes today—a roundness similar to that enjoyed by previous seasons of human experience, such as the Enlightenment or the Romantic movements. They may have been glorious or tragic, it doesn't really matter. What matters is that these seasons had an internal coherence, a harmonious design, a clear purpose, and a sense of necessity.

A meaning.

So, we now know at least this. We are not living in a civilization that came about by chance. It had a genesis that can be reconstructed, and it moved in a direction that was logical in its own way. We are not the detritus of blind productive processes. We have a History, and we are part of a Story. Of rebellion.

I can already hear the objection: OK, thanks, nice theory, but this attempt to transform Silicon Valley into a haven of libertarian revolutionaries with an awareness of their role in history sounds like a fairy tale you invented to cheer us all up. Apart from all these theories, is there anything substantial, any real facts, to prove them?

Since I was the first to come up with this objection, I'm well prepared to answer. I have a story to tell; there is no theory this time, just facts. Listen up. I'll make it as quick as possible.

Stanford University, June 12, 2005. Under the blazing sun, in a stadium packed with people, Steve Jobs delivered his graduation speech, which was later considered his spiritual testament. He closed his speech with what was to become a myth: "Stay hungry. Stay foolish." As he himself explained that day, the words were not his own. They came from a book he said was the Bible of his generation, a kind of Google thirty-five years before Google. It was a pretty weird book called *Whole Earth Catalog*: a prodigious list of objects and tools that might come in handy if you wanted to live freely and independently on Planet Earth. The catalog of things that you could learn to do, find, or buy was truly mind-blowing: you could learn to knit your own sweater or how to use a Hewlett-Packard computer, to build a geodesic dome or use drugs responsibly; there were the first mountain bikes in history alongside advice on how to grow organic vegetables, suggestions for books on female masturbation, and textbooks for holding natural burials together with news about early synthesizers. The only link between all these disparate things was Californian counterculture, which grew out of the beat generation, was honed by the hippies, and was then picked up by groups of nerds hiding out in college IT labs. This was the soil Steve Jobs was raised in (he kept the book on his bedside table). It was the SOIL MOST OF THE DIGITAL INSURRECTION WAS RAISED IN. How can we be sure? Well, hold on a little longer, and I'll tell you.

The *Whole Earth Catalog* was put together by a man named Stewart Brand. He was the kind of guy who wore fringed deerskin jackets and went around taking pictures of Native Americans. He lived in the Bay Area, had a degree in biology, was a self-declared LSD user, and was highly motivated to change the world, if there was any chance of doing so. Counterculture, as I said. One thing that may seem weird to us now, but at the time was probably considered completely normal, was that he spent a lot of time in the IT labs of Californian colleges and companies. It's not that he was a film extra or a party crasher. He was actually one of the protagonists of that world. In the mythology of the digital insurrection there is a famous session held in the 1968 San Francisco Joint Computer Conference, where the inventor, Douglas Engelbart, exhibited the first mouse for computers, the first video conference call, the first word-processing software, and the first interactive computer. This formidable Engelbart had an assistant with him that day. Who do you think it was? Stewart Brand, of course, who went on to become the first person to theorize (in between LSD trips, one imagines) the digital insurrection as a tool for liberation and collective revolt. He claimed that computers gave "personal power" back to the people; he predicted, with considerable foresight, that cyberspace would be a promised land, where communities would form a parallel world, like a fantastic version of a hippy community. He went as far as to coin an expression in 1974 that had no meaning whatsoever at the time and might well have turned out to be a monumental piece of bullshit: *personal computer*. He had seen it all, or most of it. The fact that Brand was Steve Job's hero links Apple to Californian counterculture, but this aspect is less important than what this story will eventually teach us. Brand was just the tip of the iceberg; below him there was a substratum that had experimented with the idea that developing software was a way of going against the system. In this sense, programming was no different from taking LSD or practicing free love in a VW van. Well, maybe it was a little more comfortable. For Europeans, this is hard to imagine because software developers are an organic part of the system when they are not chess pieces in the game of power. In Europe, a brother-in-law who programs computers is not going to be a revolutionary. But in California at that time, a new habitat had come about where software engineers thought of themselves as hackers, grew their hair long, took drugs, and hated the system. Just take my word for it. At that time, in that place, out of ten youngsters who wanted to turn the tables, five marched against the Vietnam War, three withdrew from life and lived in a commune, and two spent their nights in college IT departments inventing video games. This book attempts to explore the effect the latter two categories had on the world.

This is why I can conclude with some certainty that yes, it was an insurrection; yes, it was digital; and yes, they knew it. Turning the tables was exactly what they wanted. I know it's really hard nowadays to see Zuckerberg as a defender of liberty.

We are not talking about 2020, however; we are talking about the dawn of a new age. Now we know that this new age was triggered by a lucid idea of rebellion. Not everyone, perhaps, was totally aware of the social consequences of what they were doing, but most of them truly despised the system and strove to dig the earth out from under its feet. Adopting a strategy that many of them may not have been aware of, but that the sharpest of them certainly recognized, they did so with surprising determination. It was brilliantly summarized by one of them—guess who? Stewart Brand, again. He summed it up in three lines that should have been the epigraph for this book: "Many people try to change human nature, but it's a complete waste of time. You can't change human nature; what you can do is change the tools they use, change their techniques. Only then will you change civilization."

Strike!

Objection refused.

One last thing. Bending down and observing the lower vertebrae of the digital insurrection close up, we can see one more fossil—the last one, but so important it must feature in this first mappa mundi. It presents itself as a minute cluster, almost a chemical reaction: THE FUSION OF HUMANS WITH MACHINES. The choice was made with absolute clarity and cold purpose. There was not only the willingness to run the risk of becoming artificial, but also the awareness that there would be no escape without an artificial extension of our natural skills. We owe this drastic choice to the founding fathers, the early pioneers. They were the ones who were not afraid to crystallize in no time at all a posture that must have seemed unnatural at the time, but that held the promise and the hope that it would demolish the fortress of twentieth-century culture. They were the ones who designed the *human-keyboard-screen* logo. It is doubtful whether they would have done so if among the group there had been a majority of humanist thinkers. In a certain sense, it was a forced step created by the domination of minds that had been trained in engineering, IT, and science. The fact that their disciplines were coldly objective, added to their obtuse insensitivity to the seductions of human emotions, certainly played its role. These factors, put together, generated the conditions for veering so suddenly toward a pact with machines. One of our tasks today is to understand whether this choice was advantageous or not. We will get to that soon.

For now, though, let's go back to 1997. Armed with the first mappa mundi, let's move forward and discover what really happened after the game table had been established. Let's not forget that, at that point in history, the digital insurrection had only just come out of hiding and was still evolving. It had involved a tiny minority of human beings for many years, most of whom were tinkering in garages, college departments, or esoteric dot-com companies. The scenarios that developed from their inventions were sophisticated but isolated. These were the years when newspapers were opening their first online editions, calling them—rather pathetically from today's viewpoint—"telematic newspapers." Everyone had their heels on the running block, ready for the pistol, but most people had no clue what they were doing there. What would happen after the pistol went off was something that very few gambling houses would have taken money on. The issue was, would the insurrection be suffocated by the overwhelming power of traditional institutions and elites, or would it continue to burrow under the skin of the world until it caused a landslide? It is consoling to know that we are now able to reconstruct exactly what took place. As the next chapter will take pleasure in demonstrating.

Time for a musical interlude.

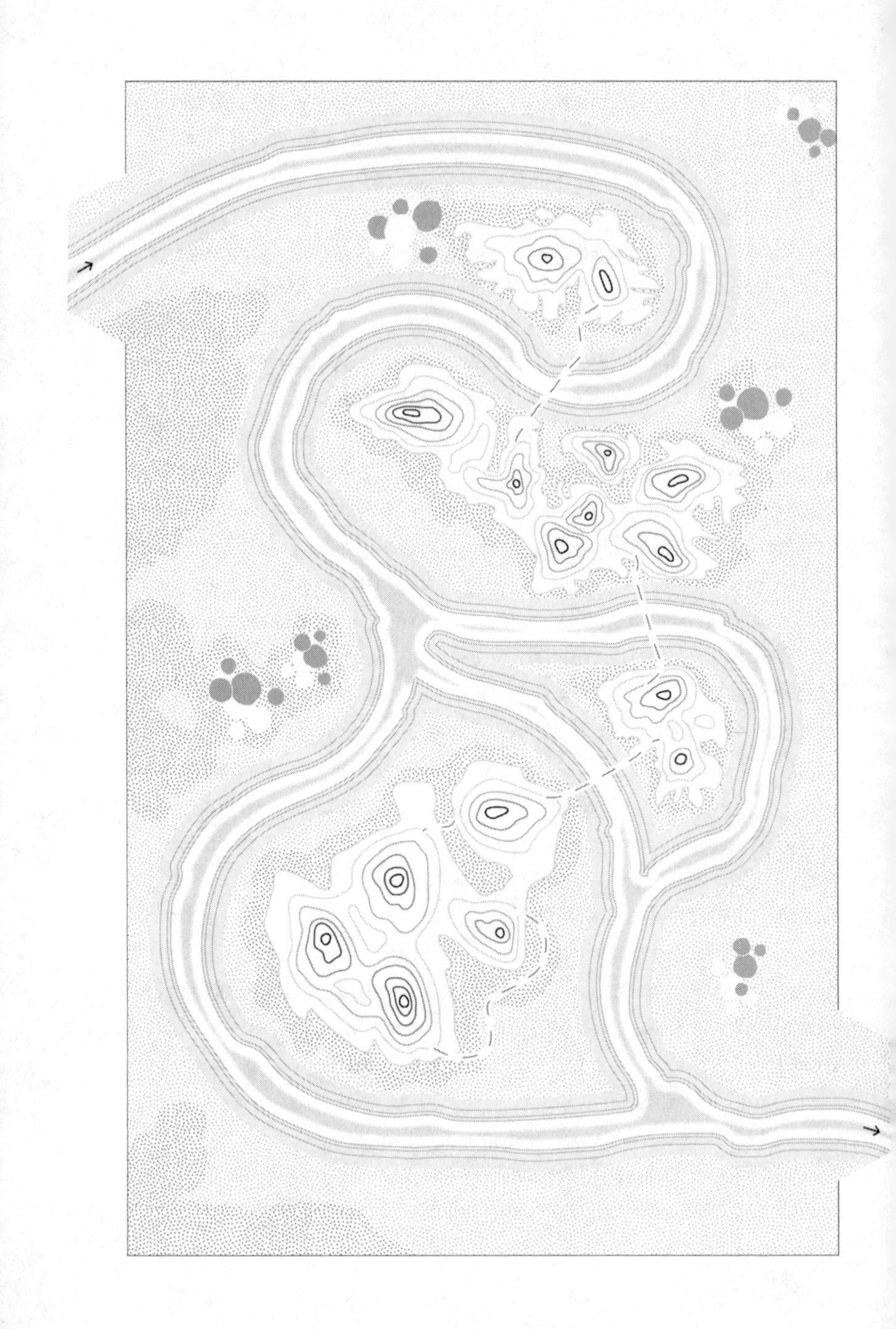

1999–2007.
FROM NAPSTER TO THE IPHONE

THE COLONIZATION

Conquering the Web

There was a new game table, and all the important pieces were lined up in the right squares. The question was, would people play? Here are a few statistics to get you going:

- At kickoff, there were 188 million internet users, corresponding to roughly 3.1 percent of the world population
- There were 2.4 million websites
- Amazon had 1.5 million customers
- 35 percent of Americans possessed a home computer

Let's leave these figures aside for now. We're going to move a few years forward, and later we'll come back to these numbers.

Are you ready? Get set, go!

1999

An American nineteen-year-old was playing around with his uncle's computer and, after a few months of tinkering, he came up with some software that did something rather remarkable: anyone could send any music they had on their computer to anyone else who owned a computer, and they in turn could share their music with you. And all of this was free. All of a sudden, the idea of spending money to buy music became old school.

The nineteen-year-old was named Shawn Fanning; the software was called Napster. Within two years, it was outlawed, but the damage was already done. Napster had instantly become a household name, and Shawn had ended up on the cover of *Time*. In the collective imagination, he had set a remarkable precedent. Basically, he taught the world that if you were clever enough and followed Berners-Lee's precepts (connecting the contents of all of your desk drawers), you could cause a heck of a lot of damage. For example, at the age of nineteen, you could destroy an entire industry—in this case, the recording labels. An essential element was, as always, total contempt for elites, which unfortunately included, in this case, musicians and songwriters. Let's just say that Napster was a lesson in how far the more radical segments of the digital insurrection could go. At the same time, it was a demonstration of the kind of extreme, unconditioned freedom it could unleash.

2000–01

The dot-com bubble burst. Financial speculation—where vast amounts of money were invested in early digital tech companies which promised shareholders immense dividends by unleashing the potential of the internet—came to a peak and finally crashed in 2001. Half of the money went up in smoke for the simple reason that, when push came to shove, most of these companies were selling things that people weren't prepared to buy. Did the whole world wake up one morning and suddenly realize this? Not really. The first ones to notice began pulling out in 1997, but the house of cards started to collapse in 2000, and dot-com shares were in freefall after that. By the end of the onslaught, 52 percent of US dot-com companies were out of business. The ones that survived were still under attack: an $87 share in Amazon dropped to $7. I can just imagine Bezos Sr.'s phone call to his son.

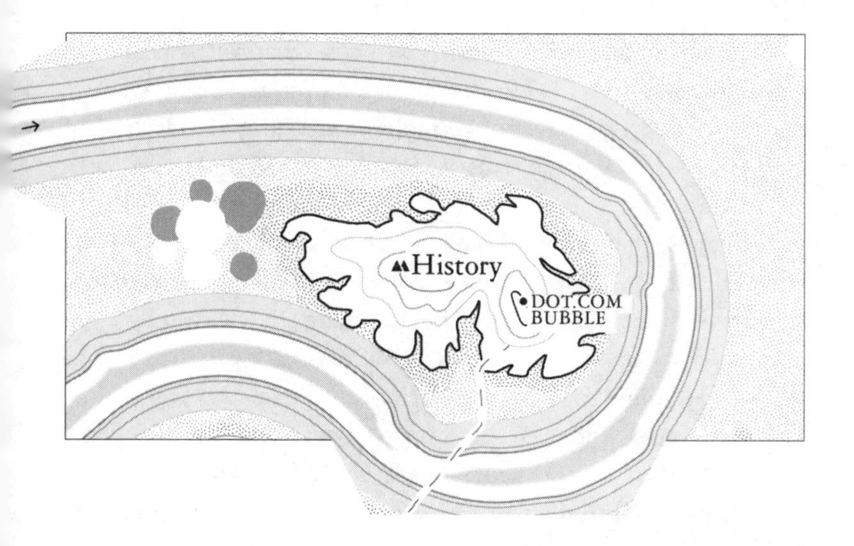

By every token the message was unequivocal: back to square one, playtime is over. Let's return to the real economy of the pre-digital age. And yet, in the light of what actually happened, one can interpret the crash differently. For one thing, when the world witnessed the collapse of the dot-com house of cards, they realized, for the first time, that it EXISTED and that it was much bigger than they had thought. This was a real eye-opener. What was significant, at the time, was that one particular species of human being—the species that wakes up every morning with the sole purpose of making money—had believed in the digital revolution to the extent that they had lost their habitual clear-headedness and had randomly gambled their chips at the game table. When these guys sacrifice their lucidity and throw themselves overenthusiastically into a bet, you can be sure something is going on. In addition, like all storms, this one swept away a lot of dry branches and left only the stronger trees standing, even though the roots were not very deep. For good or for bad, this pruning was to be providential.

You can find a list of dot-com companies that were swept away by the landslide on Wikipedia. It's an interesting graveyard, a kind of Spoon River of epitaphs to all those digital dreams. I examined the list, because I thought I might find traces of the mountains that never made it past the earth's crust or signs of vertebrae that never developed into a backbone. I took a look, and then I couldn't stop. There were some fantastic stories. Graveyards seldom disappoint. Here are a few examples.

KOZMO.COM. Based in New York, its promise was to deliver your shopping to your home within the hour, using bikes, buses, and even the subway! They survived three years.

INKTOMI. The name was a word in Native American Lakota. I had read about this search engine in Brin and Page's biography. Apparently, a driving force behind their invention of Google was that existing search engines were so bad. Inktomi, for example, didn't even list itself. If you typed "Inktomi," it didn't appear! I would have just taken it as a given; Brin and Page went on to invent Google. Inktomi was bought up by Yahoo! soon after the storm for 235 million dollars, after the company's share value crashed from 37 billion dollars.

PETS.COM. The company sold pet food online. Why would anyone want to buy dog biscuits online? Their answer was that dogs couldn't go out and buy them themselves. After two years, they went under.

RITMOTECA.COM. Some ideas were simply too advanced for the times, and they were doomed for this reason. This company was practically the first to sell music online. They were past the finishing post three years ahead of iTunes—just to give you an idea. They were based in Miami and specialized in Latin American music, but they also had Madonna and U2 in their library. Then Napster came along, offering everything for free, and the company folded, just like that.

EXCITE. This company was also a pioneer in the search-engine field. The portal was launched in 1995 by a group of students who had received four million dollars in funding (those were the days!). It worked very well, but they didn't manage to make any money. They filed for bankruptcy in 2001. A couple of years before that, two students had come by and offered them a million dollars for the search engine they had developed. Maybe as a result of the stupid name they had come up with (Google), the Excite team laughed their heads off and kicked them out of their headquarters.

OK, OK. I'll stop. You can see, though, how much fun it is wandering around the dot-com graveyard reading all those headstones.

It's surprising what you have to give up if you decide to focus on writing a book.

OK. Forget it.

2001

September 11: the Twin Towers, and all the rest. Naturally, and for many different reasons, the attack was a terrible blow for the digital insurrection, too. The most evident consequence was that it jeopardized the outlook for peace, which was the original aim of the insurrection. First the dot-com crash, then 9/11. It was a horrific roll of the dice. The fact that the attack sent an important message in such a traumatic fashion, however, should not be underestimated. The message was that national borders did not make you any safer, there was no visible front, and the very concept of war—the set of phenomena we refer to using that word—was no longer viable. (What is terrorism? If at the Bataclan theater in Paris there had been French citizens shooting rather than foreign fighters, would it have been called a Civil War?)

Thus, at the end of the day, 9/11 was, among many other

things, a traumatic, shocking, and unforgettable lesson that visibly demonstrated the constitutional foundations of the digital insurrection. That is, we need to get used to playing games in an open playing field, where certain rules are allowed, but where there are no borders. We finally grasped the fact that if the concept of war had become this fluid, then the football season was also at risk.

In light of this observation, it is interesting to see the reaction of the US government at the time. What did they do? They went and got themselves a good old-fashioned war with borders to attack and a visible enemy to exterminate. The Iraq War against Saddam Hussein, in all its destructive senselessness, can be read today as an emblem of a kind of primitive reaction to the new digital civilization. They ducked the new rules and played the same old game.

You see this kind of behavior all around you. Perhaps even within yourself. It's bizarre, because it blends together a large spoonful of dignity and pride with a small dose of absurdity. It reminds me of those soccer players rejoicing after scoring a goal after the game is already over, but the players simply haven't heard the whistle. Their faces betray that combination of happiness and solitude… they are inside a story of their own making for what seems like an eternity. They are both heroes and clowns.

Every morning when schools open their gates to their students, for example, a goal is being scored after the game is over. I think we all know that, right?

• Wikipedia, the first online encyclopedia, was launched. This was a great example of a digital otherworld being created day by day by its users, apparently cutting out all mediators and traditional elites. In theory—and to a great extent in practice—anybody can contribute, modify, or translate an entry. How can total chaos be avoided? The underlying principle is that if a clutch of well-intentioned academics put their heads together to write an entry on, say, Italy, they would not necessarily produce a better result than the one the entire population of the planet would come up with if they were free to do so. The incredible thing is that this is pretty much the case. On the other hand, it must be pointed out that the same principle underlies universal suffrage in democracy. We can't afford to question either.

The founders of Wikipedia were two American white men in their early thirties. One of the two was called Larry Sanger, and

he should be remembered for one rare feature: he was one of the very few makers of the digital revolution who had studied liberal arts. He graduated in philosophy, with a thesis on Descartes. Nearly all the heroes of the revolution were engineers. I know, it's scary. But that is nothing compared to an even more troubling statistic I am sorry to have to provide here. That is, of all the protagonists of the digital revolution, only one—as far as I am aware—was a woman. All others engaged in the epic battle were men. (It's incredible, I know. Not even in Westerns are all the main characters men!) The other inventor of Wikipedia—the one who bet his money on the startup—had studied economics. They ended up arguing and eventually fell out, of course.

2002

LinkedIn. The founder of LinkedIn should be recorded as being the first person to envisage the concept of a SOCIAL NETWORK. His name was Reid Hoffman, and he grew up in California, where he studied something vaguely related to the liberal arts: epistemology and cognitive science. The first time he thought of using the web to connect people was to solve a problem that I, personally, have never had. That is: to find a golf partner in his neighborhood. It was 1997. Five years later, he launched LinkedIn, which put people in certain jobs into contact with other people who were job hunting. In the context of this book, it was an important milestone; it was the first time that human beings produced a digital copy of themselves and posted it in the otherworld. This action, as you all know, lent itself to incredible developments.

• This was the year when the balance was tipped: 50 percent + 1 percent of data produced on the planet was now digital. Again, I have no real way of knowing this, and I'm not even completely sure I know what it means exactly (what do we mean by *data*?), but even if we take this statistic as an urban legend, the fact that the balance was tipped that year must have some significance, right? It would be nice to believe that, in 2001, the digital insurrection won a majority vote and went to power. It feels like a useful watershed to note. Let's use it.

2003

- The BlackBerry Quark went onto the market. It was a historic moment: for the first time, anybody could own and use a smartphone. Well, maybe not anyone, but certainly the forward-thinkers. It wasn't a telephone as much as a personal computer you could put in your pocket. You could also use it to make phone calls, of course, but that wasn't the point. The point was that in that tiny device, the *human-keyboard-screen* posture shifted, for the first time, away from the fixed point of a home computer, attached itself to a human hand, and became mobile.

Try to go back to those days of innocence and appreciate what a giant leap forward had been taken. It meant that wherever you

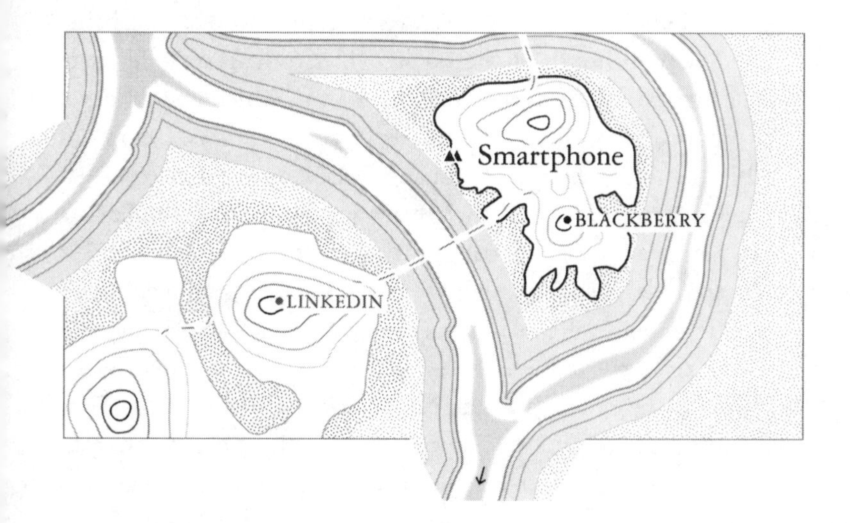

were, you could be connected 24/7 to the digital otherworld. The ones who went overboard, and actually stayed connected 24/7, behaved like drug addicts (which was when the word "crackberry" was coined). As a matter of fact, they probably didn't use their blackberries any more than we use our smartphones today. I don't know… it's hard to remember what it was like back then.

Two images stay in my mind from that historical moment: one from Tokyo, and one from New York. Both of them gave me a sense that a new age was dawning, given that in the provinces of the Roman Empire we were a little behind the times. In Tokyo, I remember, there were thousands of teenage girls walking around the streets holding their cell phones in their hands. They didn't take them out of their pockets and put them back in again; they clutched them in their hands, as if they were a fan—an analogy that immediately sprang to mind based on a cultural cliché—or as if they were a chain-smoker's eternal cigarette or a pair of glasses. They weren't a tool; they were a prosthesis. They were not a MEDIATION; THEY WERE AN EXTENSION OF THEMSELVES. Young girls, who had probably read one-millionth of the books I had read in my life, were teaching me that an anthropological shift, which I could never have imagined, was taking place before my eyes. And how did they demonstrate this? By touching the keys of the phones that they held in their hands with their thumbs (thumbs!) and doing so while laughing, chatting, eating, or smoking. They did it compulsively. They were the new logo of the digital revolution, and there they were, licking ice-cream cones at the same time. In New York, by contrast, there was a young Italian graphic artist who had become a genuine New Yorker. He was one of those types who were always a quarter of an hour ahead of everyone else, and who already sported an impressive architecture of beard and

> moustache later adopted by hipsters. Anyway, he designed great book covers. At a certain point in our conversation, he took his BlackBerry out and I looked at him with distaste. "How can you live without it?" he asked. I'll never forget the look of seraphic superiority I gave him, shaking my head. Firmly convinced of my atavistic wisdom, I bent down to look at the contraption as if I were checking a urine sample. The keyboard looked as if it had been made for tiny people: elves or fairies. The screen was equally tiny, but this guy, who designed fantastic covers for beautiful paper books, read Tolstoy on the thing while he traveled on the subway. There were so many things that needed to be understood at that moment. The fact that I remember it perfectly tells me that I didn't learn anything at the time, but that for some reason I squirreled it away, knowing full well that one day I would be culturally equipped to go back to it and read what I had to learn into it. Well, that's just what I did.

The BlackBerry died in 2016. It couldn't keep up with the revolution that it had contributed to triggering: it was a kind of Gorbachev of smartphones.

- Skype. Just as cell phones started behaving like personal computers, somebody found a way to convert personal computers into telephones that you could use without spending a cent. An interesting detail: one of the two entrepreneurs who launched Skype was Swedish and the other was Danish. From a technical point of view, the startup was prepared in Estonia. This is one of the incredibly rare cases where old Europe managed to join the rich parade of American inventors and entrepreneurs. The last time, if you remember, was ten years before, with the invention of MP3.

• One year before Facebook, its progenitor, Myspace, was launched. This was the definitive landing of human beings in the otherworld. First, it was merchandizing and information that traveled; then, it was money that circulated; then, a few fairy stories and parallel worlds were added to the mixture. Now, humans were the ones to move. Literally. It's not that they took a tour of the place, like in a video game; they actually moved into that world beyond. An example may help you appreciate what I'm saying. The singer, Adele—a phenomenon who has sold more than a hundred million discs—started by recording three of her own songs at the age of nineteen, which her friends later posted on Myspace. They were an instant hit. In this world, she didn't exist yet; in the otherworld created by this new social network, she was already a star. At one point, an independent British record label, XL Recordings, wanted to engage her. She thought it was a joke. She simply hadn't realized that the otherworld and the real world were part of the same reality with a dual pump. Moving between one and the other at the time still aroused incredulity and suspicion…

2004

• On February 4, Facebook was born. At first, it was a social network for college students, but in 2006 it opened up to any human being over the age of fourteen with an email address. Today, there are over two billion Facebook users. This was perhaps the biggest phenomenon of colonization ever recorded. Just as an example, one in two Italians today regularly disembarks into the otherworld using the ships offered by Facebook. A mass exodus, there is no other way of putting it. It will be great to go into more detail in the "Commentaries" section to see if there was any meaning to the exodus or whether it was unequivocal evidence of mass madness.

- Flickr, which was simply a social media network for posting photographs, was launched. What is interesting, in this case, is that people went to the otherworld not to live there themselves—with their faces, life stories, and idle chat—but with their gazes. Their winning gazes, to be precise. All of them achieved by means of another extension of the self that is a camera. It was a highly refined self-representation (would you go to a party having sent in advance your very best portrait photos?). Flickr did not enjoy the same level of success as Facebook. However, it launched a new technique of colonization that we still find today on Instagram and Snapchat, with interesting mental implications.

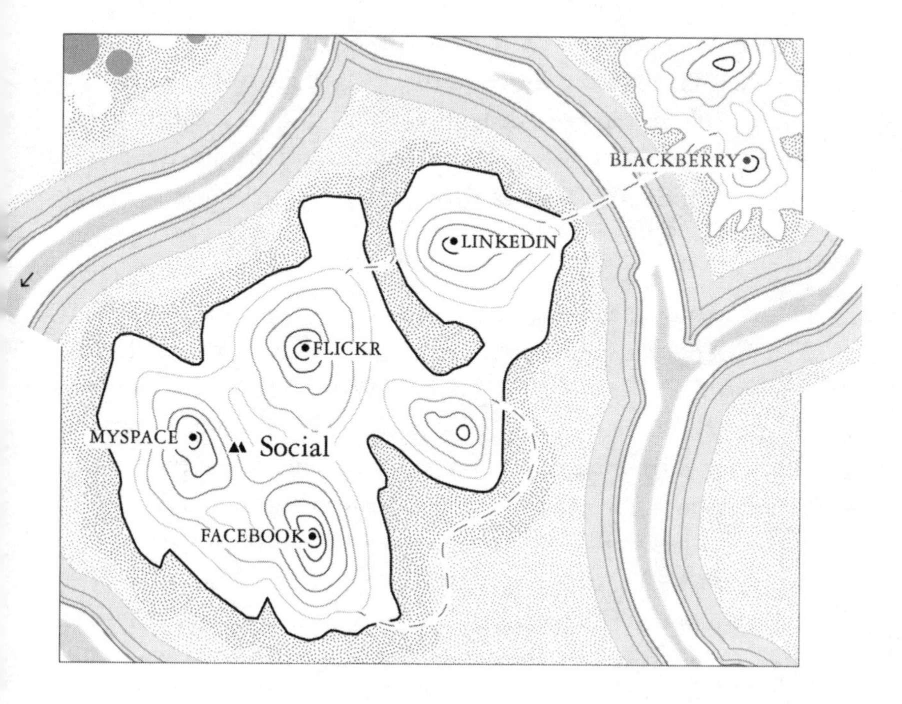

At this point, I need to add a detail. You will appreciate the paradox, I'm sure. Here it is: one of the founders of Flickr was called Caterina Fake, the only woman, as far as I know, who features in the list of inventors and entrepreneurs who triggered the digital insurrection. The only one. (The journal *Wired* reported that she wasn't allowed to watch TV as a child and spent all her time writing poems and listening to classical music. Nice, eh? Unless, of course, her surname was a coded message for imbeciles like me.)

• An Irish editor named Tim O'Reilly, who had studied Classics in his youth, coined the expression "Web 2.0." The intention was to distinguish the first phase of the web, when users were more passive (they consulted the web and surfed the otherworld, but everything was already prepared for them), from a second phase, when interactivity became widespread. In this 2.0 phase, users were called upon directly to create the otherworld. It was a useful distinction, marking a divide that helped us realize what this digital colonization entailed: we did not limit ourselves to taking possession of the otherworld; we all started to cultivate it, redesign it, and build on it. This is what Tim O'Reilly understood back in 2003.

• On September 22, ABC broadcast the first episode of *Lost*. More than twenty million Americans were watching. It wasn't the first successful TV series. *The Sopranos*, for example, came out in 1999. I chose *Lost* because it probably represents the moment that this narrative form emerged, never again to sink below the surface. The reason we are talking about it here is that TV series are an interesting example of old media (television) marrying new media (computers). Their planetary success cannot be explained without recourse to the genetic code of the digital insurrection. TV series are its highest artistic expression, which is why we have put them in this list. This is also why we should analyze them in more detail, which we will do later. For now, let's go on to YouTube.

2005

• YouTube, today the second most popular website in the world, was born. Four hundred hours of video are currently uploaded every minute. If you try to visualize a figure of this dimension, you can imagine a long line of human beings distilling their experiences into video sequences, and then transferring and storing them in the otherworld, where they can access them whenever they need or want. In this way, they contribute to generating the circular motion that has become reality: a cyclical migration of facts from

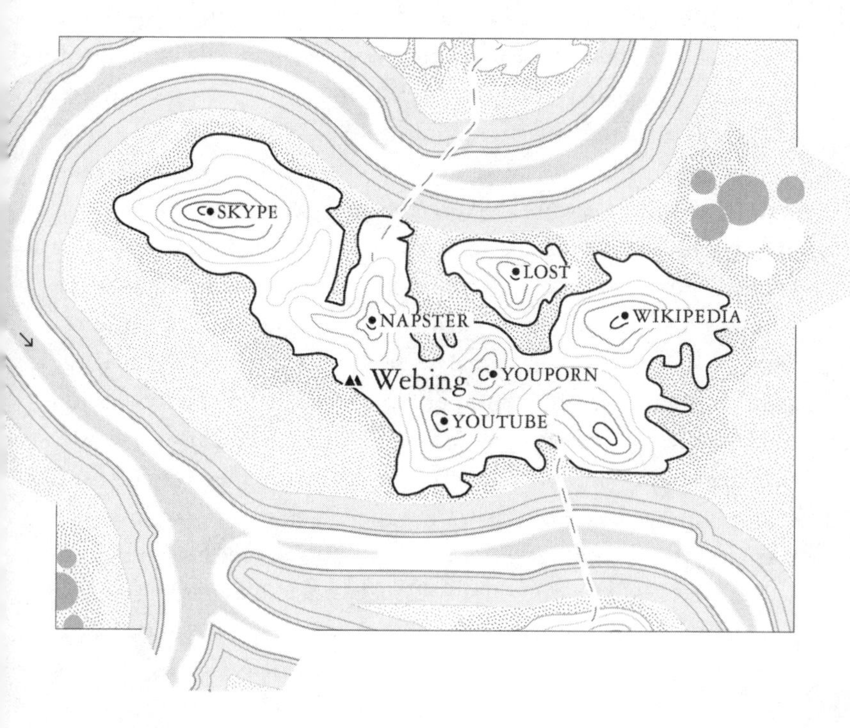

this world to the otherworld. No matter how idiotic or spectacular the substance of this action, they continue to weave their circular web, capture their prey, and call it EXPERIENCE.

2006

• Twitter was born. To appreciate what that means, we need to go back to text messages. The idea of using cell phones to write messages had been in the air for years. It sounded like pie in the sky at the time (why not use a car radio to make toast?), but the principle was actually quite sound. Back in the mid-eighties, they realized that cell phones were often unused for hours at a time, so why not use them, between one phone call and another, to send short messages in a digital format? They experimented with the technology, and it seemed to work. It was a simple matter of creating digital packages that were compatible with the dimensions of the line, which meant that the first text messages could only be eighteen characters long. They slowly worked their way up to 160 characters. There was little motivation for going any further, because they had studied postcards and realized that 160 characters was more than enough. I'm not joking. Before the new technology became an object of mass consumption, however, several years were to pass. The first cell phone to offer a simple system for sending texts was the Nokia 2010 in 1994, but it wasn't an instant success by any means. The statistics for the first year were poignantly low: the average number of text messages sent per month on the first Nokia phones to offer the service was one. Sweet. Nevertheless, after a while, people began to notice two things: first, writing messages cost less than phoning; second, writing messages was more practical than speaking to one another. By 2006, 159 billion text messages were sent in the United States alone. This is where Twitter comes into the picture. It brought together the two new technologies that were trending at the time: text messages and social networks. Twitter was very canny, creating a platform that was user-friendly, fast, and catchy. It was an instant worldwide success. At the time, the thing that caught everyone's attention—and the scorn of many—was the fact that the messages could only be 140 characters long. The limit was similar to that imposed by text messages, and people were perfectly used to texting, but the usual culprits decried it as definitive proof that a cultural apocalypse was taking place. A new humanity wanted to express its thoughts in 140 characters.

Barbarians.

Allow me to register here that today, the day I wrote these lines, President Trump—the emperor of the planet—communicated in a tweet that China secretly supports North Korea, placing the world peace at serious risk. I think you will agree that the problem is not the fact that he managed to say it in 140 characters. The problem is clearly something else. That is, that we have gotten to the point where the president of the United States communicates by using the same tool that my mechanic uses to comment on Juventus soccer matches. I must have missed something along the way. In the "Commentaries" I'll have to come back to it.

- YouPorn was born. Well, you all know what that is.

2007

- Amazon launched Kindle, an e-book reader that promised to do away with paper books. This threshold was symbolically very significant. Books made of paper were—and still are—a kind of totemic fortress in the battle between digital insurrection and twentieth-century civilization. It was a vital frontier. Bezos had, of course, weighed in with his distribution network, but he was not the first to try the same trick. In the year 2000, for example, Stephen King "published" his new book, *Riding the Bullet,* online only; you had to download it to read it. He sold it for $2.50

a copy to start with and then he gave it away free. In the first twenty-four hours, four hundred thousand copies of the book were downloaded (maybe people just wanted to see whether it could be done). Before the Kindle, Sony had also aggressively marketed an e-book reader in 2004 called the Librie—an object specially made for reading e-books thanks to a patent they had registered for electronic ink—but the fact that nobody remembers the name must mean something.

> If you are interested in how things went, here are a few facts relating to the United States, where e-books have been most successful. Since 2007, sales have never actually caught up with paper book sales. In 2011, e-books sold almost as many copies as hardback books, considering the new titles of the season. A year later, Kindle sales were higher than hardback ones, and for the next three years their sales figures soared, leaving hardback sales trailing behind. This was the time when everyone started wondering whether real books were going to disappear for good. Nowadays, people are less worried because in 2016, e-book sales slumped below the figures for hardback sales. The latter had made an unexpected recovery, which nobody commented on in particular—not even the old guard who had cried murder when e-book sales were winning the race. It's hard to understand why.

- Grand Finale. On January 9, Steve Jobs went onstage at the San Francisco Moscone Center and told the world he had reinvented the telephone. He showed the rapt audience a small, slim, elegantly simple object about the size of a cigarette case. The world soon learned to call it by name: iPhone.

Placed side by side with other smartphones on the market, it clearly belonged to the planet of *Space Invaders* rather than to that of table soccer. It was a couple of generations ahead of the game, and there was no doubt in anyone's mind that it was the product of brains that had rethought everything from scratch, leaving previous habits and logic by the wayside. Just looking at it was enough, without even switching it on. Other smartphones had a host of tiny keys that stared at you with a grin. The iPhone had one reassuring little round key, smack in the center, at the bottom of the phone. Other smartphones were minicomputers boasting their prowess. The iPhone was a computer that pretended to be a toy. Needless to say, the ploy was a great success.

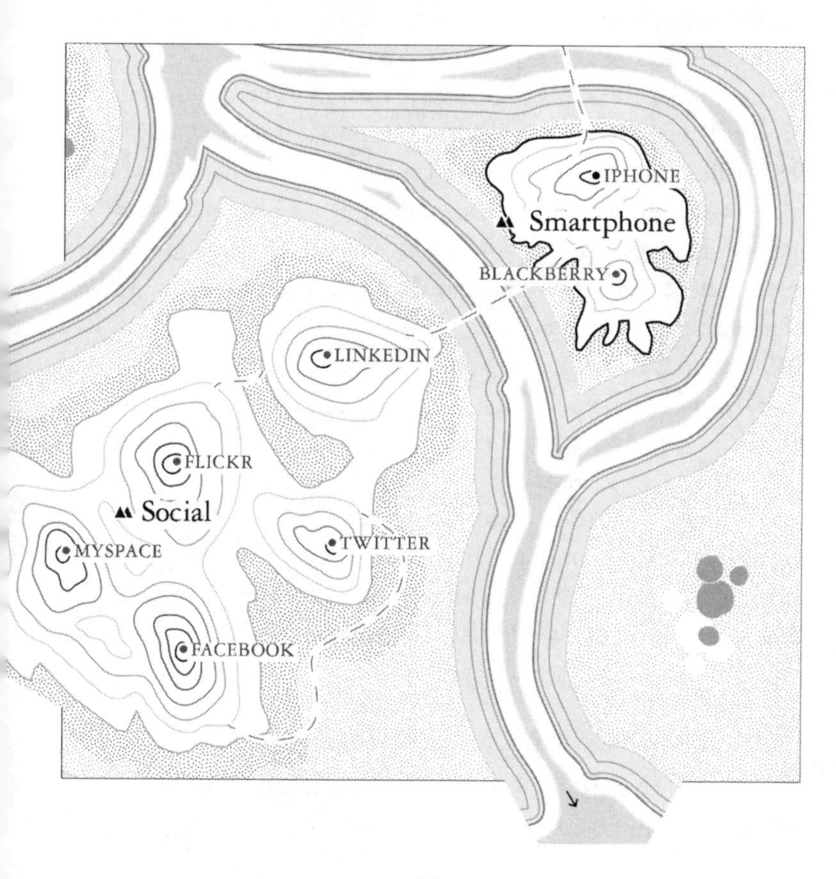

One of the things that made people's jaws drop was touch-screen technology. There was no cursor or mouse, no directional arrows, no keyboard. With a finger on the screen, you could move things, open programs, or drag objects from one place to another. There was a keyboard, of course, but it only appeared when you needed it. In any case, they weren't real keys; they were simply letters that you touched with your fingers or thumbs (the lightness of *Space Invaders* again). The technology was irresistible—as pleasurable as eating with your hands—and old Jobs knew it. You need to go back and watch that presentation to appreciate the pleasure he clearly felt as he swiped the screen with his finger, as if he were caressing butterfly wings, in front of an enthusiastic crowd. Nowadays, it looks and feels completely normal. That day, though, when he opened his contact list and, with a flick of his finger—as if he were swatting a fly off the screen—the list started smoothly scrolling up and then, like a marble rolling more and more slowly, came to a halt at the name he had been looking for... well, at that precise moment, there was a collective shiver in the crowd, a childlike wonder. Some were unable to suppress a gasp. He was just looking at his contact list! When, about ten minutes later in the presentation, he placed his thumb and index finger together on the screen and slowly opened them up, showing the audience how to zoom in on a picture, the house literally exploded. It was evident that something big was happening in that conference hall. It was like a declaration of peace between human beings and machines, a definitive transformation of the artificial into the real. Something had melted, and there was a new mildness in the air that encouraged machines to become extensions of human minds and bodies.

A few years later, when my family had already surrendered to a company that was able to charge up to fifty dollars for a charger and iPhones were familiar household objects, I happened to witness a scene that I then discovered was quite common. This makes me feel it might be useful to describe it here. My son was a little guy, all of three years old, and he had climbed onto a chair to look at a newspaper I had left on the table. He wasn't planning to read it (he wasn't a genius), but he had been attracted by a picture of a soccer player and had clambered up there to take a better look. I was watching from close by, just to make sure he didn't fall off the chair. He didn't fall; he started to stroke the picture with his finger, just as Jobs had done that day at the presentation in front of so many people. He stroked it once, twice, three times. I saw his frustration that nothing was happening. Without giving up hope, he started trying to zoom, exactly like Jobs again, with his thumb and index finger opening up gently, but nothing happened. He stared at the picture, puzzled by the fixity of the image, and I realized he was measuring the destruction of an entire civilization: my own. I knew for certain at that moment that he would never read newspapers and that he would die of boredom at school. It must be added that, given my family's northern Italian tradition, we had brought him up to believe in the value of determination, to pursue that insane habit of solving problems. For this reason, my son did not give up without one last-ditch attempt that seemed to me to express a memorable combination of rationality and poetry: he turned the page and checked out the back of the photo to see what might be wrong with it or whether there was something he could unlock. Who knows? A function that needed activating? A battery that needed changing?

On the back there was an article about the Italian basketball team.

I watched him as he climbed down from the chair with the expression of a piano bar player at closing time. I'm not sure whether I've rendered the expression correctly. What I mean is the face of the pianist when he says goodnight to the cleaner, shrugs himself into his coat, and goes home. I can't really describe it any better than that.

More or less in the same period, a friend, who had spent a period of time in California making films, came back to Italy on vacation. He landed at Milan's Malpensa airport early in the morning and went to get a ticket for the car park or for a bus, I can't remember which. Whatever it was, it was one of those ticket machines where you put money in, and the ticket comes out. I wasn't actually there, but he told me the story because he said it had "taught [him] a great deal, though [he didn't] know exactly what." Anyway, there he was in front of the machine, a little hazy after a bad night. He had been living in California for a few years, as I said, and he was young and quite smart. He did his shopping online, for example. He started touching the screen with his finger because there were a few icon-like images there. He kept on prodding the image that seemed the most appropriate until a middle-aged couple came up to him with an air of mild condescension. My friend had never seen them before but when he told me the story, he told me they must have been from the nondescript town of Cologno Monzese where they probably owned a Yarn Barn and kept the TV on 24/7 tuned to the state channel Rai 1. Whoever they were, they politely came forward and, with great spirit of collaboration, they pointed out to my friend that there were some keys on the machine and that these were the keys he should tap, not the screen. They explained this very courteously, enunciating every word carefully and—my friend told me—glancing up every now and again at the baseball cap he had on his head, as if that might explain everything.

In the end, they got the ticket for him.

I can just see the older couple in their car on their way home, shaking their heads without making any comment.

The contamination between civilizations is mysterious. The uneven step of human intelligence cannot be judged.

I'd like you to know that, personally, when I go into an Apple store and see all those people smiling at me from the Genius Bar, I cramp up on the spot. I also consider all software updates a kind of blackmail and interpret the relentlessly irritating attempts to make me buy the latest model of iPhone as an assault against my person. However, I must admit something very important, without reservation. The first iPhone was a telephone; a system for getting onto the internet; a gate to the web; a tool for writing emails and text messages; a video-game console; a camera; a vast container for music, as well as a potentially infinite number of applications from the weather to the stock exchange. Like the *Space Invader* cabinet, it potentially contained infinity, but the iPhone was immeasurably more beautiful. It sat in your pocket and weighed no more than your glasses. It officially marked the dawn of an era where transitioning to the otherworld had become an action that was liquid, totally natural, and potentially permanent. By making the *human-keyboard-screen* posture ultralight and untethering it from any kind of immobility, the iPhone imposed itself forever as a way of being—as a privileged gateway to the system of reality, the heart with a dual pump—which the Classical Era had envisaged, and which was now becoming the cradle of human experience. It achieved all this because it brought with it a mental inflexion that would turn out to be essential: IT WAS FUN. It was like a game. It had been designed for childlike adults by adult children. In this respect, as we shall see in the "Commentaries," it collected together and brought to fulfilment a distant legacy that was not just a product of the Apple mentality. The whole digital insurrection carried within it the unexpressed desire that experience could one day become rounded, smooth, attractive, and comfortable. Not reward for effort. Rather, the consequence of a game.

FINAL SCREENSHOT

If you take a good look at the backbone as we've reconstructed it so far, everything looks in place. Once the Classical Era was over, civilization advanced steadily in the same direction as before. People could have stopped, turned back, repented, or simply gotten lost, but they didn't. Like in a video game, they marched forward in a constant attempt to reach the next level without interrupting the match. Every now and again, they met their deaths, but again, like in a video game, they had more than one life to spend. The dot-com crash and 9/11 were two mortal strikes: one knocked lots of pieces off the board; the other threatened the game table itself, which represented the peaceful space required to play the game. It could have been the end of everything. And yet, it wasn't. Because they somehow stuck the game board together and started playing with the remaining pieces. They were really quite stubborn.

The results can be seen in the statistics. Do you remember the list at the beginning of the chapter? Well, let's see how things were beginning to shape up.

- Internet users in the world. There had been 188 million, the equivalent of 3.1 percent of the population. Ten years later, there were 1.5 billion, the equivalent of 23 percent of the population.
- Websites. There had been 2.4 million. Ten years later, there were 172 million.
- Amazon customers. There had been 1.5 million. Ten years later, there were approximately 88 million.

- Percentage of Americans with a personal computer. It had been 35 percent. Ten years later, it was 72 percent.

Well, the picture is quite clear, right?

Over and above the statistics, however, what is clear is that there was an almost overwhelming, collective, apparently contented inertia. We can now say with some degree of certainty that, in the era of colonization, the humans inhabiting the planet—that is, us—behaved in an entirely linear fashion: they simply expanded the game that the previous era had already started playing. They were particularly successful in two of these expansions: with social networks and with smartphones. The emblems of the decade were Facebook and Twitter on one hand, BlackBerries and iPhones on the other. They were just tools per se, but as Stewart Brand said, if you change your tools, you'll build a new civilization. In fact, these two tools carried within them two seismic shifts—if you'll allow me the expression—that were destined to change society. Let us see how.

ONE:

Social networks certified the PHYSICAL colonization of the otherworld. What I mean is, people PHYSICALLY moved there. They didn't only move their files there; they actually relocated themselves, their profiles, their personalities. Or, in more sophisticated cases, such as Flickr, they moved their reverberations, the warmth of their emotions, the vibrations of their desires, the world that they wanted. At the same time, people transferred to the otherworld an increasingly large share of their social relations. If you measure the distance between texting a friend and posting a tweet that tens of thousands of people may read, you will have a good idea of what took place over a very short interval. We webbed ourselves, linking all of our desk drawers like Professor Berners-Lee did; we decided to communicate in the same way information was communicated on the web; we found in the otherworld a system with no attrition that allowed us to fling every action and word out into the open sea of a community that appeared to have no limits or borders.

Don't misunderstand me. I'm not saying that WE WENT TO LIVE IN THE OTHERWORLD. We colonized it. It's different. We linked it up to the world and started pumping blood quite efficiently into the dual-pump system we invented with the web. Social networks made everything extremely clear. Nobody actually called the movers and relocated to the otherworld (apart from some total nerds, who may have). Most people simply learned to circulate their personalities on two different circuits, which were ultimately twin hearts of a single organism—reality. Whatever people say, we have become quite expert at the game. Thirteen-year-old kids are perfectly familiar with going back and forth constantly between the two worlds. The idea that there might be a border between the two would probably seem an inappropriate way of describing their experience. They live happily in a home where there is a system of reality with a dual pump. Experience, for them, flows through a circulatory system with twin hearts. Asking them which heart beats where would be gratuitous. The question tackles an important issue, but it should probably be formulated in a less childish manner.

TWO:

The process of mass colonization of the otherworld, and the physical relocation of people over the new border, was definitely accelerated by the second emblem of the era: smartphones. The shift was visible. By removing all the inflexibilities of the *human-keyboard-screen*

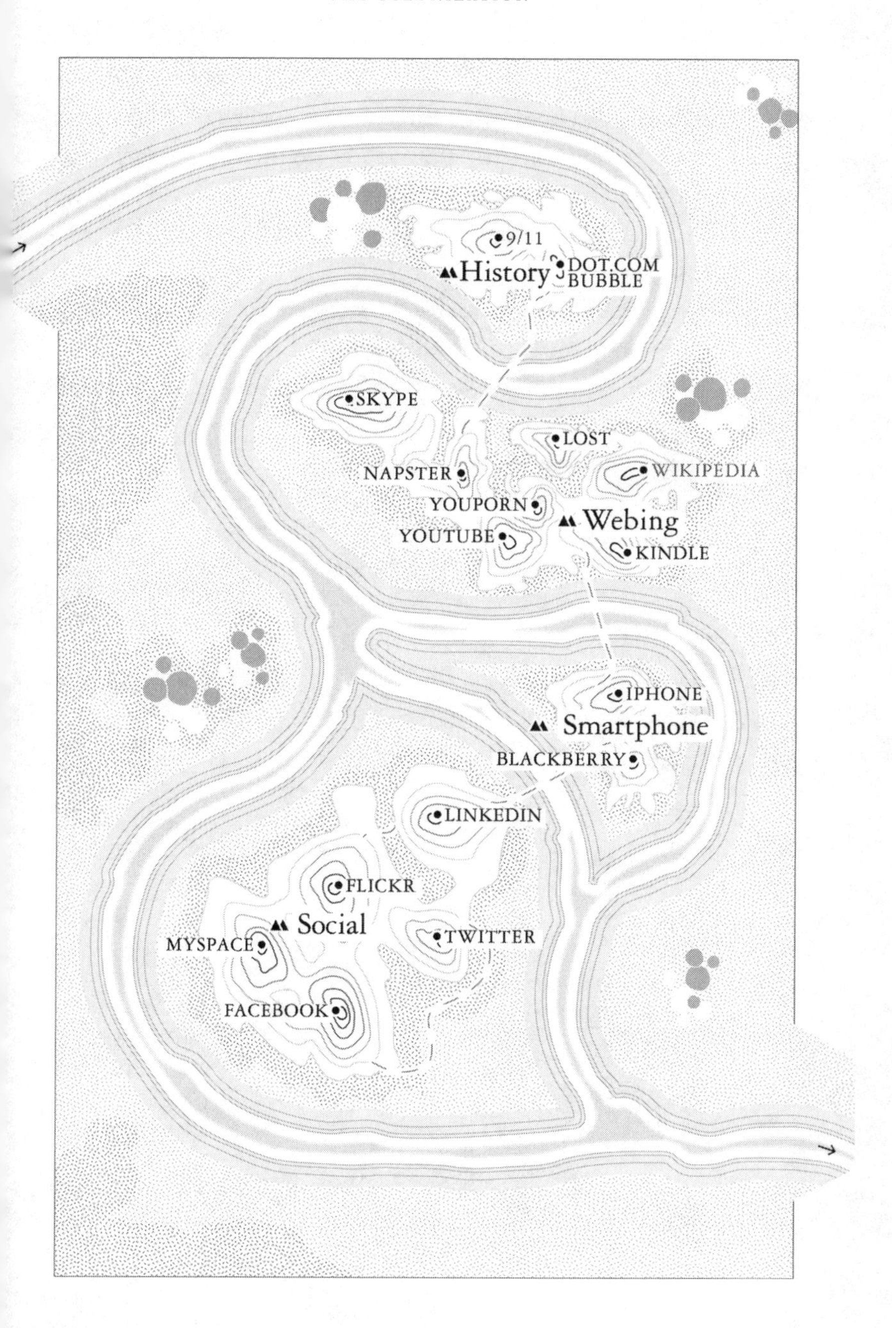
9/11
History
DOT.COM
BUBBLE
SKYPE
LOST
NAPSTER
WIKIPEDIA
YOUPORN
Webing
YOUTUBE
KINDLE
IPHONE
Smartphone
BLACKBERRY
LINKEDIN
FLICKR
Social
TWITTER
MYSPACE
FACEBOOK

posture, migration between the world and the otherworld was made much easier. There was clear continuity with the intuitions of the previous era: both the decision to adopt the posture, and the invention of the otherworld, expected technology to come up with a design to share them with the rest of the world. That was exactly what happened.

Thus, roughly thirty years after *Space Invaders*, we have a clearly outlined mountain range; a landscape we can legitimately start to call a new civilization. Not an electrifying technological turning point, but an actual civilization.

The surprising—or should I say problematic—things we may find when we dig down and examine the fossils will become clear in the "Commentaries" that follow.

COMMENTARIES ON THE ERA OF COLONIZATION

THE GAME

I went back and watched the video of Steve Jobs presenting the iPhone again. I wanted to examine it more closely, dig down, and find fossils. There was something to discover, something that would take us far. In the end, I convinced myself that what made it unique was the fact that Jobs was having SUCH A GREAT TIME. I'm not saying he was enjoying standing up there on stage being cool; what I mean is that the iPhone itself gave him extreme pleasure. He wasn't taking pleasure in TALKING about it; he enjoyed USING it. Everything about his manner was aimed to communicate the idea that iPhones were FUN. I know this may sound totally predictable to you, but you need to go back to that time to appreciate the message. You need to go back to life before, to what we came from. Was the disc phone with a receiver FUN? No. Was a public phone FUN? No. Was the BlackBerry FUN? Well, not really. They were all tools that solved certain problems, but nobody ever thought they were supposed to be FUN, and so they weren't made to be fun.

The iPhone was fun, which is what Jobs was desperately trying to communicate in his presentation.

HE'S SAYING IT'S A GAME.

So, there's the clue we're looking for.

Try to go back in your mind and remember how many times in your life you found yourself in a situation where the solution to one of your practical problems WAS A GAME. There can't have been that many. I bet the few you could come up with were back in the distant past when you were a child, because the early pioneers of "gamification" have always been parents: feeding you by making the fork into an airplane, providing you with a potty in the shape of a spaceship. Dad turning into a monster, a kite, or a cactus according to what problem you were relying on him to solve. I used to open my son's diapers in the spirit of a panner in the Yukon Valley searching for gold nuggets (I once actually found a coin in there). Anyway, what I meant was that the iPhone was designed to solve a great many problems, but it did so with the idea of being a fork-airplane. Everything about the device was there to remind you constantly of the fact; its solutions looked and felt like childhood games. The colors, the design, the icons that looked like candy, the font that cool kids would use, the single button (games for toddlers usually have at least two). Touch technology itself was obviously childish. What do you think the middle-aged couple from Cologno Monzese thought of my friend when he stabbed at the screen on the airport ticket machine? They most likely thought he was an overgrown baby (an impression confirmed by the baseball hat, of course).

Games and childhood, then. Let's not fall into the trap of thinking it was just a matter of packaging, design, and appearance, though. Jobs's candid pleasure up on that stage suggested something more substantial. The iPhone—like its forebearers, the Mac and the iPod—did not just look like a game; in many ways it was a game. IT WAS CONCEPTUALLY DESIGNED AS A VIDEO GAME. With an object like that in your hand, what do you end up doing without even realizing it? Adopting the *human-keyboard-screen* posture, made more pleasurable by the touch screen, you got your little tasks over and done with so that they weren't hanging over you: calling your mom or looking for a restaurant. Small problems called for modest countermeasures, all of which were pleasurable to the touch and to the eye, as well as being accompanied by satisfying sound effects. Do you need to call Gigi? Touch four keys and you're done. Do you want to take a picture of Marisa? Three taps and you're done. Do you want to get rid of that photo, because Marisa had her eyes closed? Two swipes and you're done. And so on. For more expert players, there were higher levels of the game: go onto the web, buy music online, write an email. Even then, it was a game of call-and-response: a little Martian falls on your head, so you kill it. Whatever you needed to do, that device was not really a phone, and it wasn't even really a tool; it had every appearance of being a video game. Scratch that. A whole raft of video games rolled into one.

What I later discovered was that this did not happen by chance or because Jobs had dreamed it up. It was a long evolutionary process. Video games were one of the foundation myths of the digital insurrection, after all. They were one of the highest divinities up on Mt. Olympus, revered by those people. I'm not saying this because I happened to base my reconstruction on *Space Invaders*. I'm saying it because, historically, video games were the cradle of many later inventions developed by the protagonists of the insurrection. Would you like to hear two stories that explain a great deal?

The first one regards the legendary Stewart Brand, the one who first said, "Stay hungry, stay foolish." Well, in an interview in the British daily *The Guardian* a few years ago, he described how things started back then in California and talked about the people he had met who had opened his mind. At one point, this is what he said: "I was at the Stanford computation center, and this was some time in the early sixties, and I saw these young men playing *Spacewar!* They were out of their bodies in this game that they'd created out of nothing. It was the only way to describe it. They were having an out-of-body experience, and up until that time the only out-of-body experiences I'd seen were drugs."

Of course, the only thing you'll remember about this little anecdote is the link between video games and drugs, and if you have any children of your own, this is probably one of your worst nightmares. But I must ask you here to leave all that aside, not to get distracted, and to listen carefully. About ten years after having that illuminating experience, Stewart Brand wrote a long article for *Rolling Stone* which made history as one of the earliest, most brilliant, and most prophetic theorizations of how computers would change our lives. This article was the first time that somebody actually put into words what must have seemed like the ravings of a lunatic at the time: the prediction that every human being would one day own a computer, which they would keep on a desk at home; that computers are a source of power that should be distributed; and that they would make life easy, peaceful, and reasonably happy. It was a historic article, believe me. His style was not very sophisticated, if I may be so bold, but the contents were a bombshell. What do you think the name of the article was?

Spacewar!

The name of a video game.

Half of the piece was about the video game, in fact. Why? Brand says it was because *Spacewar!* was the perfect crystal ball, one that allowed you to see where computer science and the use of computers would lead.

Is it clear by now that these guys had video games in their DNA? That they weren't just dickheads who thought the world was a game? Rather, it was that video games were their starting point, and this fact left its mark forever.

If you still harbor any doubts, listen to the second story.

Story number two is about Steve Jobs again. In 1983, he was

invited to speak at a conference of designers in Aspen, Colorado. I'm not sure how well-known he was back then. He was there to try to help the conference attendees grasp a few fundamentals. I do know that most people knew zilch about computers in those days. At one point, when he realized that not one person—not one!—knew what software was, he tried to explain. As an aid, he compared software to TV. In so many words, what he said was that a TV program can replicate experience: if you watch JFK's funeral, you get emotional when you relive the experience. By contrast, he went on, if you launch software, you are not capturing experience, but rather THE UNDERLYING PRINCIPLES OF EXPERIENCE. The designers had no clue what he was talking about (neither have I, for that matter), so he said, "Don't worry; I have a perfect example that will help you understand what a computer is." What was this perfect example?

A video game, of course.

He chose the game *Pong*. Does anyone remember it? It was a rudimentary ping-pong match that could drive you to distraction, which came out in 1972, six years before *Space Invaders*. Anyway, he started talking about *Pong*. In order to explain to a crowd of outsiders who had no idea what computers were, he chose an example that in his mind most faithfully embodied the unprecedented, revolutionary capacity of computers: a video game where you had to hit a ping-pong ball.

It is thus not that surprising that twenty-four years later, there was Jobs presenting the iPhone with such amusement, giving the impression he was holding a video game. He did have a game in his hand; we know this now without a doubt. And yet, he had always been holding a game; he had never held anything but games: throughout his life as a hacker, he had been playing games like *Pong*. Ultimately, this would not be very significant if it were not for the fact that he was just one of many. He may have been a little more aware than others, but he was still one of many. Video games were the gym where most of the hackers who triggered the digital insurrection had trained. They were somehow the mental pattern that summarized most clearly the vague intuitions of those scrambled brains. They were seeking a world, and instinctively they envisaged it as having the design and logical structure of a video game.

This inclination was replicated almost every time a solution was sought for a problem, which could only have led, in the long run, to creations like today's smartphones or environments like Spotify and Tinder. Basically games. Already in the Era of Colonization, more than ten years before, the trend had been quite evident. Google was a game but did not know it yet (in fact, you wouldn't really call it FUN). Facebook had a clear game component, with a comfortable, convenient, amusing interface. Numbers of "likes" and "followers" are up on screen as if they were points in a video game, recovered and metabolized with incredible ease. Twitter learned the lesson by becoming a tool that basically shoots match after match with the results bouncing back immediately (retweets, likes, etc.) in an entertaining, uninterrupted stream of winners and losers. In the meantime, web page links continued to offer attractive, glossy panoramas of glaciers in the otherworld. Napster played cops and robbers, emoticons took texting by storm, and Kindle attempted to market itself as a magic whiteboard. All these without even citing actual video games, which were relegated, like hidden viruses, inside other devices. Games had been elevated to the status of a founding myth of an entire civilization. Since that time, living had become an intricate series of matches, where the harshness of reality was the playing field and the emotion of experience the trophy. In many ways, it was the land promised by the hackers: a single, free, nonstop video game. *The Game*.

* * *

I don't think I need to stress any further the fact that we are at a crucial stage right now (says he, stressing it). It's a fact that many of our fears and doubts—which are by no means unfounded—have been triggered by this idea of the Game. At a certain point, we began to notice with some irritation that many of our actions no longer responded to the slow, sensible pace we had learned all our lives. All of a sudden, our movements were required to be brusque and totally lacking in poetry. Perhaps we have finally hit upon a possible genesis of the phenomenon: the world as it is today was designed by the people who invented *Space Invaders*, not table soccer.

I once asked a friend—who was no fool—why he insisted on buying vinyl records, the old 33¹/₃ rpms.

Rather than firing off the usual excuses—such as that the sound is better, etc., etc., etc.—he said, "Because I like getting up from the sofa, going to put a record on, and then going back to the sofa." He really loved music, and what he was telling me was that listening to music was so important to him that he somehow valued the slow, somewhat strenuous, slightly ceremonial action required to put the record on the turntable. If you are wondering how we went from a civilization that valued this kind of refinement to one that came up with Spotify (where you can change a record with a click), now at least we have a partial answer: because we chose the path that led to the Game. I'll be a little brutal here: for historical—possibly Darwinian—reasons, from a certain point in time (since the iPhone, I would say if I had to hazard a date), there was no chance of anything surviving if it didn't share the same DNA as video games. I'll even compile a useful checklist of the required genetic characteristics of the species needed to survive:

- An attractive design that gives sensory satisfaction;
- A structure that can be reduced to the repetition of an elementary problem/solution pattern;

- A minimum time lapse between the problem and its solution;
- A progressive escalation in the difficulty of the game;
- No need and no use for being immobile;
- A learning curve provided by the game itself, not by studying abstract instructions;
- Immediate accessibility, without any explanation;
- The reassuring presence of a point system after a certain number of steps.

That's about all I can think of for now, but I have some important news for you: if you are engaged in an activity that doesn't have at least half of these features, you are doing something that has been dead in the water for some time.

You are allowed to be a little unsettled by this news.

SUPERFICIALITY
Reverse Thinking

At the iPhone presentation, if you go back and watch the video, there was one word that recurred almost obsessively. One word.

Simple. Very simple. Very, very simple.

Simple.

Whether you wanted to listen to a song by the Beatles, phone a friend, go onto the web, turn the volume up, or switch the whole thing off—all that was required on the iPhone was a tiny action that was not only fun, as Jobs kept reminding his audience, it was also simple. Very simple.

It may sound like something completely obvious, without any significant consequences. Nonetheless…

Simple is not just the opposite of *difficult*. It is also—above all, in this case—the opposite of *complex*. What Jobs really wanted to reveal to his audience was that the iPhone allowed highly complex processes to converge into the absolute purity of a simple action. He wasn't saying he had *simplified* the telephone. Not at all. What he was keen to point out was that he had created an extremely complex tool that was extremely *simple* to use. Somehow, the device had managed to conceal all its complexity in a hidden compartment. All that was left, floating on the surface, was the pared-down fruit of the complex processes; the final synthesis; the rudimentary, practical heart: icons to touch, lists and pages to scroll. Eyes on the screen, fingers tapping: the sensation one had was that our actions had been cleansed of any dross; everything had been reduced to an ultimate form of simplicity; anything indispensable had risen to the surface, while everything else had been sucked down into an invisible non-place.

The sensation was very pleasant, and those friendly, smiling, colored icons epitomized it perfectly. Now it is easier to see how their babyish appearance masked something far more sophisticated: they were the emerging tips of immense, highly complex icebergs hiding somewhere behind the surface of the screen. Ironically, the icons adopted a stylized image of the tools they were at that moment destroying: a telephone receiver, a compass needle, an envelope, a clock with hands. There was even a cog. Destined to disappear as objects, they were transformed into buoys that marked the precise point where the practical heart of things had risen out of the sea, leaving behind the complexity of twentieth-century processes that had been holding them down. They were there to mark the fact that THE ESSENCE OF EXPERIENCE HAD EMERGED FROM ITS UNDERGROUND CAVES AND CHOSEN THE SURFACE AS ITS NATURAL HABITAT.

I want you to know that we've finally reached the heart of the digital culture.

At the end of the day, it was just a phone, you might say. You're right, but when you're dealing with people who want to change human minds by changing their tools, you have to be wary of how they made those tools. We must be clearheaded enough to recognize that the iPhone created a mental framework that was destined to have an enormous influence on the way we live in the world. The image of an iceberg is easy to recognize: a colossal mass of complexity invisible beneath the water level; a miniscule, indispensable heart floating on the surface. Highly articulated mathematical operations, stored in underwater warehouses, generated elementary results that could be easily read on the clear surface. Hard work was abandoned in a time *before*, while experience was on offer as an immediate, spontaneous gesture.

An iceberg.

OK. Pay attention here, because this is a crucial point. The interesting thing about this mental framework—the iceberg—is that IF YOU TURN IT UPSIDE DOWN, YOU GET THE MENTAL FRAMEWORK THAT DOMINATED TWENTIETH-CENTURY CULTURE.

I grew up with that twentieth-century framework in my head, so I know how to describe it. On the surface, floating right under your nose, was chaos or—in a best-case scenario—the treacherous trap of *superficial* perceptions. The game at the time was to get past them with the help of teachers selected for that purpose. Following a path that needed hard work, application, and patience, what was required was to go down *in depth* where, like an inverse pyramid, the complex articulation of reality would slowly reveal itself, first in the clarity of a few elements, and later in the blinding epilogue of true essence. This deep core was where the AUTHENTIC MEANING OF THINGS was preserved. We used to call the moment we gained access to the hallowed hall EXPERIENCE. It was rare, and almost impossible, without some form of mediation: from high priests such as professors, from books or travel, or sometimes from suffering. Whatever it was, it implied dedication and sacrifice. The idea that it might be a game, or that it could be *simple*, was not contemplated. In this sense, EXPERIENCE was seen as a rare luxury or as a reward for the privileged few. In any case, it was always the legacy of a caste of high priests. It was, nonetheless, a glittering prize that was highly sought after in the weary emptiness of our lives.

As you must now appreciate, the figure of the upside-down iceberg is equally clear. We applied it to all the various rivulets of our existence. Whether we were finding out about a piece of news, trying to understand a poem, or attempting to build a relationship: the framework was the same. It was a reverse pyramid. At the top, we had to move quickly: the surface was powdery and full of crevasses. Then, slowly and patiently, we worked our way down and attempted to reach the essence of things. Hard work on top, the reward below. The image is pretty clear, right?

Turn it back the other way, please.

What do you see?

An iPhone.

The prize is on the top; the hard work is below. The essence is on the surface; all the complexity is hidden somewhere else.

The iPhone is just an example. Wasn't the first Google page, a white sheet with only about twenty words to explain everything, the tip of an iceberg just like an iPhone? What about Berners-Lee's twenty-one words on his first web page? Or the Windows 95 screen with its comforting, orderly line of icons and pre-established commands? Weren't they the same? They are all iceberg tips. Behind them, under them, inside them—we don't know—there was a mass of complexity, but the essence of things was floating on the surface. You could locate them with a glance, you could understand what to do on the spot, you could access them immediately (without mediations, without the help of high priests). The iPhone is made this way, as is Google, Amazon, Facebook, YouTube, Spotify, and WhatsApp. They present a kind of simplicity where the immense complexity of reality rises to the surface, cleansed of any dross that might weigh down its essential core. The result is a concise catalogue of existence that would have comforted Aristotle, delighted Darwin, and excited Hegel. All of these philosophers sought essence behind appearance, simplicity behind complexity, the principle before the multiple, the synthesis after the difference. I'm sure they would have appreciated the YouPorn homepage, if they'd had time to waste on that kind of activity, that is.

Now we know that, with tools of this kind, the digital insurrection struck twentieth-century culture with a mortal blow. It disintegrated its basic principle that the kernel of experience was buried in the depths, and that it could only be reached by dint of hard work and with the assistance of some kind of high priest. The digital insurrection grabbed that kernel from the talons of the elites and brought it out into the open. It didn't destroy, annul, banalize, or miserably simplify it: EXPERIENCE WAS LIBERATED ONTO THE SURFACE OF THE WORLD.

We are now in a position to state that these human beings who reversed the previous order rejected the myth of in-depth knowledge and instinctively closed the gap between appearance and essence. In their view, APPEARANCE AND ESSENCE WERE THE SAME THING. What they wanted to do was funnel experience into basic elements that could be arranged on a desktop and accessed with simple, quick actions. They were driven by a fear that should not be underestimated: the fear that the heart of things might once again sink into immobility and that access to the kernel of knowledge might once again be taken over by a caste of high priests. They had seen the catastrophes brought about by that framework, and they instinctively chose solutions that made it impossible to return to such a hellish state of affairs. They had a strategy in mind that was, in its way, brilliant: if there were such a thing as authentic meaning, they needed to make sure it was not isolated; they needed to bring it up to the surface and make it visible. The only way to make sure it wouldn't be a monolithic secret, sanctioned by the same old powers that be, was to guarantee it became the product of various currents of existence, of the transparent and changing footprints of human endeavor. That is, impermanent.

People of this kind developed technologies that were suited to their way of thinking. They weren't philosophers; they were mostly engineers. They didn't develop theoretical systems; they built tools. Their reverse thinking led to actions, solutions, and habits. Insignificant actions (checking the weather or measuring one's temperature), when they were multiplied over and over again, generated a mental posture that was not a random effect of successful tools, but rather the coherent reflex of the reverse thinking that originated them. In the long run, what happened is that we ended up expecting from life what we saw working in our trivial daily activities. If, in order to make a phone call, I only had to tap a screen to choose a number from a restricted range of options—where a plethora of possibilities had been rendered clear, organized, and even fun—why were things so different at school? Why would I choose to travel any other way? Or eat? Why should I try to understand politics when it is so complex? Or read a newspaper? Or find out the truth? Or, perhaps, find someone to love?

Thus, little by little, we all started to think in this reverse fashion. We all started to adopt the useful rule that any game could be played as long as you could arrange your pieces on the illuminated board that is the surface of the world. While things were still hidden in the depths, supervised closely by the caste of high priests, they were more chaotic. The world was unfair, dishonest, and dangerous. In a spectacular collective effort, we started to unearth the core of the world and bring it up to the surface—we discovered the habitat we were best suited to living in. We didn't intend to strip the world of its most authentic meaning; we simply wanted to place it where it would be able to breathe more easily.

You must agree it was quite an awe-inspiring idea.

THE FIRST WAR OF RESISTANCE

It certainly was awe inspiring. However, it was also—and the time has come to remind ourselves of this—devastating. Objectively speaking, the coupling of the Game and Superficiality was hard to digest; it forced the old world to migrate in such an extreme, unexpected, and shocking way, that alarm bells began to ring, high and low. It is true that twentieth-century civilization cleaved to the great cultural and political institutions, but, as we have seen, the strategy of the insurrectionists was to circumnavigate the fortresses and head straight for another destination: the place where people selected the tools they needed for their everyday survival. That is where the progress of the Game was most sudden and almost unopposed. Moreover, in 2002 we had already gone over the edge and definitively adopted digital language. By relentlessly digging underground tunnels, the insurrectionists were beginning to cause landslides up there in the old world.

This was when twentieth-century civilization realized, for the first time, that it was under siege. It wasn't understood as much as it was felt. There was a sensation of being attacked by an invisible enemy: they could hardly see it, they didn't know where it was, and they didn't know how to fight it. And yet, they could see the dusty ruins of villages which, until the day before, had seemed destined to prosper for a good while to come. The alarm bells were periodic, protracted, almost pedantic—like an anti-aircraft artillery that carried on after the planes had already flown by. This was the season when people plastered the walls with posters defending dairies, if you remember. When I wrote *The Barbarians*.

The barrage—naturally led by the elite, who were beginning to feel the ground caving in under them—was chaotic, arrogant, and ultimately blind. There was one thing, however, that they understood. That is, there was an element of the Game that seemed to be stripping human experience of its most complex, sophisticated, or mysterious aspects, diminishing everything to a simplified system that bypassed hard work, reduced the unit weight of facts, and chose the easiest and quickest solutions. It was a vague, still unfocused intuition, but it was true that the Game, to put it crudely, appeared to be robbing the world of its soul. They felt as if it was establishing a secular, functional, game-like version of the world for people who had no particular interest in engagement.

As an exposé it was irresistible: who would want a world with no soul, designed by people who sit at the PlayStation all day? At the time, anyone who had anything to lose from the eventual success of the digital insurrection had a blazing banner to fight under: they were defending human beings, or at least a refined, noble idea of human beings. The battle was thus ratcheted up a notch, which helps us date the first significant war of resistance against digital culture: in the years between the two millennia. Since those leading the resistance strategy were mostly certain twentieth-century intellectual elites who were very unfamiliar with digital tools, the fighting took place in defense of traditional actions such as reading, eating, studying, or loving, and against aberrations such as mega bookshops, fast-food chains, mass tourism, or love in the era of YouPorn and Facebook. The old elite felt those areas in particular were heading for disaster and tried to bolster them up the way they knew how. They had no inkling at the time that everything had started long before, triggered by a new form of intelligence that was building tools suited to their dreams. Nor did they grasp the fact that their world and the otherworld were not two environments at war with each other; they were two hearts of a single system of reality. For these reasons, they were fighting their war of resistance with obsolete weapons, without understanding where the battle-front lay, using tactics that might have been strategic in a previous game, but which no longer existed. In practical terms, they were playing a video game as if it were pinball. They claimed to know who had stolen the ball and demanded it back in a macho fashion. In some particularly pathetic cases, they even debated whether the flickering light effects could be done away with.

And yet, something important came of that first war of resistance, which must be recorded, respected, taken seriously, and placed under a microscope. That is, the basic intuition that the Game was perilously stripping the world of its soul and robbing human experience of its nobility.

Was this an optical illusion? A convenient lie? A refined form of blindness? Up to a certain point—yes, it was.

Jobs was so gleeful up on that stage that day as he played around with his iPhone. In reality, however, something was about to happen that, if one thinks about it for a second, was actually quite scary. For one thing, all the elites were about to be toppled, because they weren't carrying a survival kit for living on the surface of the world. Having already lost most of their legitimation, they found themselves on the brink of extinction. When teachers fear for their future, it's never a good thing. Fear makes them biased and blindly aggressive. Nor is it a good thing when they get fed up, send everything to hell, and leave. An empty desk leaves an ambiguous message; it may feel like freedom, but it is also a step toward an emptier world. This is particularly true when the rise to the surface of a whole system of values had led to something akin to a *jailbreak*, indiscriminately freeing prisoners with new forms of collective intelligence, alongside those with old forms of individual stupidity. For a long period—that may not be over yet—only well-trained, neutral observers could distinguish between prophets and dickheads. That's enough on its own to worry stupid people and sound the alarm for smart people. The core of the world was surfacing and melting into a great Game, but not without any suffering. It didn't just rise to the surface and deny that a few things had been left behind.

Then again, even someone like me, who had instinctively welcomed the digital insurrection, had the sensation that the numbers just didn't add up. If you'll allow me a personal memory, I admit that I was mostly taken aback by the hypocrisy of those who claimed to defend the status quo but were actually defending themselves. At the same time, however, I had the feeling that we were leaving something behind. It wasn't what the elites were lamenting the loss of (mostly their privileges and profits). It was something more important, buried somewhere deep within our collective psyche like *the memory of a vibration*. It was irritating to think about, but I kept on going back to it: WE WERE LOSING THE MEMORY OF A CERTAIN VIBRATION. I don't know how else to describe it... I know it's hard to grasp the concept, but I have an example that may illustrate what I mean.

While all this was going on, I was making a movie. It was 2007, and the film world was on the brink. In those days, we shot movies with old techniques and then transformed everything into digital form for editing, mixing, and special effects. Finally, the whole thing had to be put back into cannisters, because the projectors in the movie theaters still used old-fashioned film reels. To sum up: we worked analogically, then digitally, then analogically again. It was a heck of a lot of work, and we weren't very good at using the new machines, so a lot of things got lost in the process of going back and forth between the different technologies. It was clear that things had to change. In a couple of years, film reels would be obsolete (in those days, to make a movie you shot enough film to cover a soccer field). Kodak, the main manufacturer of film, declared bankruptcy in 2012. May they rest in peace.

However, at the time, as I was saying, we were still walking the tightrope between the old and the new, and the whole thing was keenly debated. Since I was in the middle of making a movie, I decided to try to follow the debate. I thought it would be a good case study, illustrating the clash between the digital insurrection and twentieth-century civilization. It was fascinating. The conflict was harsh: the digitals marched forward, looking back scornfully at the analogicals, who were shaking their heads while shooting their last few miles of film and proclaiming that the end of cinema was nigh. It was not just a matter of taste or pixels, you see. The debate centered around the filmmaker's craft: the digital technique, they complained, changed the lighting, the weight of the movie camera, the time it took to shoot, the cost—everything. Generally speaking, it appeared to simplify things but—and it was a big but—the old artisan filmmakers knew that digital photography killed much of the beauty, the magic, what some might call the *soul of cinema*.

Which brings us to the core of the question.

We were talking about the movies, not the world, which meant I could actually go and investigate the matter. I asked my director of photography to project a scene of the movie I was making in a screening room, first from a film reel and then digitally. I wanted to see the difference—if there was a difference—with my own eyes. I wanted to see what was missing, because that would be the soul. I know it was a bit childish, but I thought it was quite a clever move, and I hope you agree.

When you are the film director you can ask for anything. They screened it, as I had requested.

And this is what I saw: *there was no difference*. The color palette, the sharpness, the depth—all identical. Of course, my director of photography, who was sitting beside me, could see some differences. It was his job. When I asked him whether a normal moviegoer would ever notice any difference, he said no without any hesitation.

Then he said: "Look at the edge." The edge of the screen. At that moment the film was screening from a reel. The edge of the screen was wavy. Not very wavy, just a little. *Like a vibration*. He went on to screen the digital version. "Look at the edge," he said.

Stationary.

Film does this, he explained, making a circle in the air with his hand as if he were cleaning a window. Digital photography doesn't. I understood that the screen looks as though it's breathing with a film reel, whereas with a digital version it is stuck on the wall, completely static.

I will always remember that circular movement of my photography director's hand, and now I know that what we miss in every digital device, and more in general in the digital world, is that breathing, that waviness, that irregularity.

Like a vibration.

Well, I know it's practically inexplicable—and if you don't know what it is, you'll never know you're missing it—but if I had to sum up what was genuine about the instinctive resistance of some human beings when they realized they had unwittingly signed up for the Game, the expression that springs to mind is *like a vibration*.

Has it gone forever? Will kids who are ten today ever know what it is? Are we heading for collective amnesia? Was that vibration what we used to call a *soul*?

It's hard to say, but if, like me, you keep on looking for answers, this is the answer that comes to me: a vibration is a movement that makes reality ring true; it is an unfocused image where reality breathes in meaning; it is a delay where reality produces mystery. It is, therefore, the only depository of real experience. There is no real experience without a vibration of this kind.

Olé.

You may well object at this point that the people who put up posters defending dairies, who dug their heels in, who fought the war of resistance, were right after all!

No, they weren't.

Let me see if I can explain.

POST-EXPERIENCE

I took me a while to get this, too. On the one hand, the digital world had stifled the vibration which I KNEW was the heart of experience, and on the other, I couldn't really swear that the world created by digital technology was dead, muted, or without meaning. It just didn't add up. You could claim it in bad faith, to defend your interests. Many people did. If you looked at things with a little naiveté, however, you could see pulsations everywhere in the Game. There was something there that still throbbed, breathed, produced experience, generated intense meaning, and kept the soul alive. It was hard to grasp where that energy came from, where the pulsing beat was hiding, but to deny its existence was foolish. The most obvious cliché was the younger generation, whose flesh the digital insurrection had already entered, and whose behavior and posture had already been altered. For those of us who came from the old civilization, the young were hard to read. It's not a good idea to generalize, but the sensation was that young people did none of the things that we considered necessary to create experience, meaning, or intensity. The way we viewed them, they were destined to be total morons. But that was not the case. There was clearly an intensity there, a meaning; moreover, they had a strength that, compared to ours at their age, seems spectacularly resilient.

Where the heck did that strength come from?

It's a little easier to understand now.

If you have a console of daily life neatly arranged in a series of basic elements where the complexity of reality is first tamed and then revealed to you in such a way that you can access it quickly (your iPhone screen), there are basically two things you can do.

The first is to use those basic elements to solve your life: most of the work is done by others, you use the elements provided, and that's that. It's like clicking on an icon. You solve your problems and save time. Period. That's all well and good, but it's clearly a fairly basic use of digital culture. Those guys turned twentieth-century culture upside down and created an iceberg, bringing meaning up to the surface for you. And what do you do? You book a restaurant online. Watch a video on YouTube. Create a WhatsApp group for your soccer team. OK.

Vibrations? Zilch.

OR YOU CAN DO SOMETHING ELSE. You can use the iceberg; use the fact that someone else has unearthed the essence of things and positioned it on the surface of the world; use the fact that you have a console full of basic elements that are easy to manage; use the fact that, whatever they put on that console, it communicates with all the others; enjoy the fact that there are no longer any high priests around to bother you. Then, you can do the only thing that the system seems to suggest: put everything in motion. Cross over.

Connect. Superimpose. Contaminate. You have cells of reality at your disposal: simply arranged and easily accessible. You don't just use them, you make them WORK for you. They are the result of a geological process, but you should start using them as chemical reagents. You can join up the dots to outline an image. You can string lights together to form whatever shape you want. You can cover enormous distances in an instant and develop geographies that didn't exist before. You can superimpose jargons which have nothing to do with one another and create a language that has never been spoken. You can transport yourself to places you have never seen and lose yourself in faraway locations. You can roll all your convictions down any downward slope and watch them as they form into ideas without knowing why. You can manipulate sounds, allowing them to explore all their potential, and then you can take on the task of recomposing them into a new resonance, which might even be beautiful. You can do the same thing with images. You can create concepts that are trajectories, harmonies that are asymmetrical, buildings that occupy space at different times. You can build and demolish, over and over again. All you need is speed, superficiality, and energy. Your way of being is in motion. You must never stay still: going down in depth slows you down; the meaning of any figure you create is linked to your ability to move with sufficient speed; you are in many different places at the same time, and this is your way of inhabiting one of them, whichever one you are looking for. If you have made things work well for you, it will not be difficult to detect in your steps a strange effect, a modification that alters the text of the world, that appears to put it in motion: LIKE A VIBRATION.

There you are, see? It's the soul. The soul is back in the picture.

I've decided to call this particular way of being POST-EXPERIENCE. It's not perfect, I know, but it'll do for now. It gives you an idea. It's the experience we have to try to imagine once it has distanced itself from its twentieth-century model. It's the experience we can achieve with the tools of the digital insurrection. It's the experience that stems from superficiality. The first time we glimpsed it was in a trivial, annoying phenomenon we called "multitasking." It was all there when someone from my generation watched a millennial do five things at the same time—all of them, in our view, badly, superficially, senselessly. What was really happening was that they were doing one thing we didn't understand, and they were doing it fantastically well. They were using the seeds of experience—which had been refined over years until they had taken on those features of completeness, essentiality, and inevitability that only seeds possess—and they were cross-fertilizing them in order to cultivate a vibration that, in the long run, would allow them the privilege of true experience. Post-experience, that is.

It's also completely possible that these multitaskers were simply neurotic kids who couldn't sit and watch TV without playing *Minecraft* at the same time. Even if this were the case, though, there was something embedded in the dynamic framework of multitasking to which digital culture owes its idea of post-experience. The fact that some of these people wasted it thoughtlessly, or were firing blanks, was part of another problem: we all have to throw our lives away somehow, right? I can assure you; we wasted our lives back in the previous century, too. However, a thousand imbecile millennials—if you can actually find them—are less relevant than the one person who was experimenting by means of multitasking the motion that (sooner or later) would afford them the opportunity to squeeze meaning out of the world. That person is the one who informs us what post-experience is.

The lesson is that the people who were signing petitions in favor of dairies were wrong. What they were registering was the fact that the methods adopted by the digital insurrection had sacrificed a certain sophistication, which meant that people felt they should defend what they considered the soul of the world. However, they were perhaps not naive, objective, or intelligent enough to realize that experience wouldn't just die like that, nor would the human passion for a certain vibration that represented the meaning of the world. In their own way—a way they had learned through constant use of their tailor-made tools—these new human beings continued to pursue something that was like a glow of intensity, like out-of-focus reality, like a vibration that mysteriously clung on to the facts, like a continuous last stab at Creation. We can now say with some certainty that they had disassembled the soul of the world, taken it down into the depths to preserve it, and then put it back together in a place where they felt it was safer to pass on to future generations. Obviously, if you went and looked for it down there—where we used to keep it—you would get the distinct impression it no longer existed for anyone, anywhere. This is a mistake we've already made. Repeating the error would not only be fatal, it would also be grotesque and, unfortunately, useless.

DISMAY

By contrast, it would be useful to devote time and intelligence to understanding what we don't know about post-experience—anything that would be useful to discover. It's hard to do, of course, when you're studying the Era of Colonization because, at the time, post-experience was still underground, indistinct, and limited. It was only in the era that followed, the Era of the Game, that it became more clear-cut and emerged explicitly as a feature of collective behavior.

Nonetheless, one thing could be intuited in the age of the iPhone, Facebook, and YouTube. Something I've had in mind since I started thinking about writing this book, and that I'm now going to try to put down on paper for the first time, because writing this book has in some way clarified things for me.

Here it is: POST-EXPERIENCE IS TIRING, HARD, SELECTIVE, AND DESTABILIZING. OK, I agree, in the same way that a video game can be tiring, hard, selective, and destabilizing. Anyone who believes the Game is an easy environment hasn't understood a thing. iPhones are easy. The Game isn't. Living in the Game isn't. WINNING the Game isn't. It's anything but a walk in the park.

I would go as far as saying this: ultimately, the main difference between the twentieth-century idea of experience and the idea of post-experience stemming from the digital insurrection does not lie only in the issue of depth versus superficiality. These are vital ingredients, of course, because they are two models that are diametrically opposed; they represent a total reversal of the previous way of thinking. Agreed. However, at the end of the day, the greatest difference is another. People in the twentieth century envisaged experience as a sense of completeness, fullness, roundness, and fulfilment. Post-experience, on the other hand, is seen as a sudden opening, exploration, loss of control, or dispersion. Experience was the conclusion of a formal action, the reassuring result of a complex operation, the final return home. Post-experience, by contrast, is the beginning of an action, opening up to exploration, a ritual of estrangement. Like TV series, which are creatures of the digital age, there is no real ending. And there is not an end in itself. It is the duration of a movement, the trajectory of the undertaking. Experience had its own stability; it communicated a feeling of solidity, of permanence in itself. Post-experience is a movement, a trace, a crossing; it communicates impermanence and volatility; it sketches figures that never begin and never end and names that update themselves continuously. Experience was linked to categories that were supposed to be clear-cut and magnificent in their constancy: truth, beauty, authenticity, humanity. Post-experience, however, is a movement, which means it cannot cultivate anything on firm ground. Sure, truth, beauty, authenticity, and humanity could still be harvested, but as part of ever-changing processes, shifting constellations that constantly regenerate themselves, relentless fluctuations between shores that are also shifting. Let me try to sum it up briefly: experience was an action; post-experience is a movement. Actions order the world; movement destabilizes it. Actions seal things up; movement rips them open. Every action is a point of arrival; every movement is a point of departure. Actions are harbors; movements are open sea. Finally, actions are firm; movement is VIBRATION.

If you are able to grasp what I'm trying to express, I can finally pin something down. You almost certainly already know it, but now you'll be able to conceptualize it better: post-experience often generates dismay. It could not be otherwise. It creates uncertainty, bewilderment, disorientation, loss of control. It is becoming our way of creating meaning, of finding the vibrations in the world, of rekindling the soul of things. However, the price is an underlying instability, a relentless sense of impermanence. This is why, despite all the predictions, the Game has turned out to be a tiring, hard, selective, and destabilizing habitat. I suppose there's always option one available: tap a few icons and solve your life; limit yourself to booking a restaurant online. And yet hardly anyone stops at that; everybody steps out on the path of post-experience sooner or later. We are all hungry for the soul. It's just that the Game starts getting hard exactly at this point. Some people drop back, others leap several squares forward, disparities are created, and ultimately something becomes clear. It was a fact that the insurrection had not reckoned with: not all human beings are equal in the Game. There are better and worse players: those who play better end up conditioning the game board, turning the table in the direction they want it, becoming the guardians of the game or its principal players. That is, they end up becoming what we can now call by its real name, though it might surprise you: the elite.

Ouch.

This has always been the case, you might object. Privilege of experience has always been the prerogative of those who adapted best to the environment or, more simply, of the rich. True, but wasn't it one of the dreams of the digital insurrection to break the chain of privilege and open up access to experience to all? How the heck have things gotten to the point that the cards have been shuffled differently, but the game is still the same as before?

In sum, one of the facts we tend not to take sufficiently into account is that the Game is a harsh environment: it provides intensity rather than stability; it generates inequality; and it is unsuited to many of the people who are nonetheless obliged to live in that environment. Moreover, most public institutions—above all, educational establishments such as schools—do not prepare for the Game. They do not hone the skills required to survive in this unforgiving habitat. At best, schools may teach you to lead your life in a brilliant, twentieth-century, post-war, democratic fashion; they certainly do not train you for the Game. It's easy to see, then, why so many people today are struggling, and why an ever-widening gap between elites and the rest of humanity has again been created. You can see why a considerable proportion of humanity has reverted to a basic use of digital tools and now devotes all their available time to shoring up all the security they can. If you are wondering why, for example, nationalism has started rearing its head again, and why there is a revival of the idea of borders—forgetting the disasters that only two generations ago threatened the world order—here is a possible explanation. Imagine finding yourself in the middle of the Game: you are sobering up after the inebriation caused by augmented humanity, and you realize you are floating in a game they never taught you to play, which you are losing and not enjoying one bit. What do you do? The only thing you can do is retreat as far back as possible, until you find a wall to lean on, which will at least protect you from the snipers behind you.

A wall, please.

We'll build a good old national border. Wouldn't you like that?

Yes, I really would.

Right, let's get building.

Is there only regression, ignorance, and egotism in this instinct to seek a protective wall, any wall at all? I really hope you don't think so. There is also—yes, also—a legitimate dismay, whose precise source we can now locate. We started creating this dismay when we turned the twentieth-century figure upside down, when we chose the surface of the world as our habitat, when we started to navigate this surface and try out a way of life that we called post-experience. We aggravated the sensation of dismay when we decided that the Game—like the iPhone, Google, or WhatsApp—required no instruction booklets, teachers, or training sessions. We then made things even worse for many, many people when we forgot to string up a security net for those who fall. In video games, after all, you always have more than one life; you can always go back to the beginning and start afresh. But we forgot the little detail of the safety net.

This is how we ended up in this mess.

Can a process of liberation disorient humans to the extent that they are ready to return willingly to their cages? Is this what is going on?

HOMELAND

Companies such as Amazon, Google, Apple, and Facebook, in the meantime, have become immeasurable, unfathomable monoliths we no longer know how to consider. What we need now is to make an effort, take a step back about ten years to the Era of Colonization, and try to understand what happened. That is when several different events cross-pollinated and began to grow, breaking the earth's surface in those very years and preparing the terrain for the environment we're living in now.

The first event was that some of the dot-com companies—quite a few of them, actually—began to make money in spades. There's no point in listing them here; they're the same old names. However, a profit hike of that kind had never been seen before, not even during the industrial revolution. The question is: were these exceptional profits the pretext for the digital insurrection? The answer is both yes and no. Amazon was designed to make money and never attempted to hide the fact. Microsoft's mission had an equally cold commercial approach. Google and Apple were already a little different. In these two cases, the urgency to make dividends for their investments went hand in hand with the pure pleasure of making a vision come true, or even making the world a better place. It would be hard to say whether their thirst for profit was more or less important than their unadulterated egotism. Zuckerberg, for example, was quick to monetize his intuition, even though it was not nearly as visionary as some of the others'. The man who invented email didn't make a penny; Wikipedia has always been nonprofit; the web (potentially the biggest money-making machine ever conceived) was literally bequeathed to the world and to whoever wanted to use it, but its inventor gained nothing from it. All in all, in that crowded cohort of insurgents there was a bit of everything: from pure prophets to

Wall Street sharks, from improbable idealists to entrepreneurial profit-seekers.

This allows us to determine that any attempt to condemn the digital insurrection as a colossal market-oriented operation would be historically unfounded and basically incorrect. There is one thing to add, however. That is, FINANCIAL RESULTS WERE COMMONLY ACCEPTED AS A VISIBLE POINTS-SYSTEM, MARKING WINNERS AND LOSERS BETWEEN OLD AND NEW. When we are talking about behavior, mental frameworks, the dissemination of practices, the easiest way to understand what is going on is to count the dollars. It simplifies things. Thus, translated into a language everyone could understand, the vertiginous commercial success of some of these companies became the yardstick for measuring success in conquering the center of the game board. It was the points-system of the video game, if you see what I mean. In fact, when I think of the way Zuckerberg, Jobs, or Brin and Page made their moves during those years, I can't help but see a pattern that transcends the traditional dynamics of capitalism and leads me right back to the video game. I can't help but see them as brilliant nerds, playing a paranoid game of their own invention; almost without competitors, they were without any real need to tear their adversaries to pieces. There they were, obsessively trying to reach the next level, exceed every previous record, push the game to its extreme. Perhaps they were unheeding of the financial profits; they were simply lost in their personal challenge, immersed in their game, devoured by neurosis. As Stewart Brand would have put it: "totally out of it."

The second event, which may have arisen from this form of self-hypnosis, is worth mentioning. This was that the original justification for the digital insurrection was almost entirely lost. The trend was to forget there had ever been an enemy (twentieth-century culture) and adopt the future, *that* very future, as a reason in and of itself. There is always a moment when victorious rebels against a system become the system, in their turn. The Era of Colonization was the moment when it dawned on the digital insurgents that they might be able to take over the government of the system, as it were. These companies didn't control that much at the time; they simply began to evolve according to their own logic, forgetting where they came from, the revolution they had fought, the fears they had combatted. They were not so much a consequence of the past as a lucid, almost fanatical invasion of the future.

While all this was going on, a small group was forming among the big players. They were laying down the weapons of their guerilla warfare and preparing to govern reality. It is hard to give this group any other name than a new elite. They were not the Silicon Valley programmers and engineers, who preferred to work behind the scenes. They were something very different. They were the ever-widening community of people who were capable of post-experience, who knew how to use a system of reality with twin pumps to travel effortlessly from this world to the otherworld, who made movement their natural habitat. In this case, too, value was assigned through numbers. A kind of aristocracy developed, nurtured by the quantities of movement they were able not only to tolerate, but also to generate. Numbers measured these movements, like the points-system of the old-fashioned video game: followers, likes, etc. It was still under the surface; there were no YouTubers at the top of the bestseller list of books, let's say. Influencers, at the time, were still less powerful than news anchors. And yet, something was changing. While one segment of humanity was just beginning to slip down the snake of dismay, another popped up out of nowhere, climbed up the ladder, and started living in what for years had been considered a promised land, and was now, amazingly, looking more and more like a homeland.

These are the years when the digital insurrection began to

settle. It didn't stop; it simply abandoned its nomadic habits, pitched its tents, and took possession of the land it had been promised. This was done with a strategic design in mind, a managerial class trained to implement it, a limited but well-tested set of rules and regulations, and a massive war kitty. The rising dismay of some segments of humanity had not yet become organized into protest, while the resistance of the old, intellectual elite had been worn down to almost nothing. On the other hand, the settlers could rely on the somewhat passive, but doubtless real, complicity of many others, who had chosen digital tools for their survival. Everything was ready. The mission of the colonial invasion needed to be completed. Action was needed in order to make sense of the journey undertaken, the perils surpassed, the courage required for the enterprise. A city needed to be founded. It already had a name: the Game. Building it was the next challenge.

The glorious days of the insurrection, as you may have realized, were coming to an end.

POST-EXPERIENCE OF ONESELF

The reason we can talk about "passive" but "real complicity" of large numbers of people was the instant and inescapable success of social media networks. We could even say that social networks were the tools the digital insurrection used to enroll a vast number of inhabitants in the Game. Needless to say, I've spent a great deal of time studying them. As before, I got on my knees and started digging down into the earth, looking for fossils as clues. I was surprised to find the excavation less interesting than I thought it would be. Maybe I'm missing something; I'm not sure. I have this feeling that the DNA of social networks did not develop there. It evolved in other places and was then left to develop in a specific ambient: that is, among people. To put it perhaps a little more clearly, the very fact that social networks existed was the natural consequence of moves that had been made elsewhere.

Once the otherworld exists, people are obviously going to go there. Once a system of reality with dual pumps exists, people are going to circulate on both sides. If post-experience is what we have described it as, people's personalities—their *authentic* personalities—are the result of the sum total of their presences in both this world and the otherworld. These moments blend together by means of chemical reactions that form a kind of ultimate identity, which is portable and unstable. Try adding up all your online presences—each one with a different profile, because being on Twitter is not the same as having a Facebook page, as you well know—and you'll find there's quite a constellation of constantly twinkling stars. Add to this what used to be called "real life"—the things you do in the world—and you will appreciate that, at any given moment, your personality is a big, wide open construction site. The term we used to express this concept was *augmented humanity*.

It's not easy to keep all these different elements of your personality in order, in fact, because, as we've been repeating, the Game is hard. Many people manage to move around perfectly in that double helix of this world and the otherworld. Many others don't. They stammer here, make a move there, post a picture of their swimming pool somewhere else, and that's it. Once again, disparities and elites are created through ranking mechanisms. We've seen it; it's the way things go. For some, augmented humanity is a way to enrich their lives, while for others, it is a hopelessly wide playing field that is both dispersive and destabilizing. And yet, we all have a go at the post-experience of OURSELVES. This means that *all of us*, whatever our skills, education, or destiny, are in this together. We are all living in a scenario we ourselves created, whose features are now clear. About ten years ago, part of that mysterious vertical pump line that was our personality rose up to the surface and settled in places where it could be observed and where it was exposed to the wind of other people's gazes. It is not the detritus that we send to the otherworld to stockpile. They are authentic pieces of our matrix, which we translate into formats that are compatible with a universal language so that we can release them and let them drift away on the stream of collective discourse. What we expect, in exchange, is to exist more fully, to be recognized, perhaps to explain ourselves better, definitely to understand ourselves better, and, ultimately, to be more visible to ourselves. The phenomenal success of digital and social media platforms reveals the prosaic fact that on our own, shrouded in the silent mystery of who we are, we won't get very far. Social networks give us witnesses; they help us exist in the light of other people's gazes; they help us carry pieces of us up to the surface; they help us speak/demonstrate/represent/give form. They help us convert bits of the mystery into self-propelling objects, which we can roll out onto the surface of the world. It's so complicated to experience oneself that using the post-experience of ourselves as an aid is a perfect solution. It's hard to criticize it as a strategy.

Some people will inevitably object that we risk paying far too much time and attention to the otherworld and not enough to this one, which deserves more care. What's the point of living your life on social media when you don't even notice someone passing by under your nose? There's no point, of course. However, once you start getting used to using your digital devices, you are on a slippery slope; you will inevitably slide down toward the theoretically more convenient otherworld. It's just the way it is. Can anything be done? Hmm. I wonder. I would, nevertheless, note that we may be missing an element that would help us focus better on the issue. It's important, so let me use an example.

I once happened to talk to two people who were much younger than me. Not only were they present on social media, they actually worked with it. You know, the kind of people who work as social media advisors for private companies or public bodies. I offered them dinner out in exchange for a chat about what they did. I wanted to see whether they could explain a few things I still couldn't grasp. I was looking for fossils to give me clues, as I told you. Well, they had a lot of interesting things to say, of course. I was delighted to hear, for example, that every social network has its own measure of the average distance from the truth they tolerate. That is, at any given time, you can choose how close to the truth you want to be. You can decide to post a photo on Instagram or write a couple of lines on Twitter. You are choosing, perhaps unconsciously, but you are still choosing. You decide whatever distance from the truth you want. At one point in the evening, we started analyzing my mechanic's Facebook page (yes, the same mechanic!). What I really wanted to know was what people hope to achieve by posting all that stuff. I wanted to observe things IN DETAIL, with them as my guides, hoping they could explain. Since I wasn't getting anywhere watching pictures of deer flash by alongside selfies of my mechanic on the snow (his life), I got impatient and started to grumble. Anyway, I asked them whether they did stuff like that, posting pictures of *deer* on their Facebook pages, and do you know what they said? They said, "Yes, of course we do." In the midst of the mental battle I was engaged in, one of the two—the woman—came out with a phrase that struck me dumb. Once, she said, at a concert of a rock band, the National, it was all so perfect that LIFE DIDN'T NEED ELABORATING. She told me she didn't tweet, she didn't send any WhatsApp messages, she didn't even take any pictures. Nothing at all. She said this as if it were extraordinary, and as she said it, I was listening, listening really carefully, and maybe I understood something I hadn't understood before. That slippery slope is not just a gameboard that is slanted downhill so that you are encouraged to take easy, digital, speedy, comfortable actions. It may well be slanted in the other direction: uphill. That is, when we bounce pieces of our life against the otherworld, what we are doing is ELABORATING that life. Thus, when we pick up our smartphones—instead of being in a place watching, listening, or touching our surroundings—it is not just the instinct of a zombie who doesn't know how to live. We may also be doing it for the opposite reason. That is, because life is never enough: we could be capable of so much more, so we go and grab all the extra we can get and send it off to the otherworld where, maybe, life will finally live up to our expectations.

That's what I thought, anyway, as she was telling me this anecdote. Exactly with this image: the slope is not necessarily downhill, it may even be uphill. We make our way up it, in order to find the post-experience of life we are looking for.

This is why I still find it unacceptable when people sitting at the table with me are chatting to someone else on their phones, and I am not convinced that all those lit-up phone screens in front of me while I am giving a lecture are for taking notes. And yet, I need to accept the idea that during the irritating back and forth between the world and the otherworld, we may be successfully dealing with our solitude, and perhaps extracting from life a brilliancy that normally only presents itself in flashes, and even then only when it feels like it. In short, there is a new skill here that borders on neurosis, often slips into stupidity, but also exists in a higher form. That is: when you use the Game cleverly, in order to give things the vibration we feel we deserve from them. When you know how to climb up the sloping game board. This is one of the forms post-experience takes. As I was born in 1958, I remember well feeling there was no remedy for the solitude and dumb boredom of life. There were some drugs that could help, but they were hard to get and excruciatingly limited. The sickness was much cannier than the cure. We made do by dreaming about worlds that didn't exist or cultivating every drop that life conceded with old-fashioned care. Don't be lulled by your memories; it was a meager harvest back then, before we were distracted by picking fruit from the otherworld. We sowed lots of seeds, but the yield was controlled. It was considered normal. There was nothing glorious about it, if I may say so. We would have bartered everything we had for a perilous journey with the promise of an El Dorado, where we had been told we'd find more light, bigger cages, and days that went by at greater speed.

Come to think of it, that is precisely what we did.

MAPPA MUNDI 2

There were a few guidelines that evolved as a result of the shocks of the twentieth century, and that stayed with the insurgents. These were: always choose movement, skip steps, cut out mediations, dematerialize experience, do not fear machines, and adopt the *human-keyboard-screen* posture. There was also a clear method that derived from the kind of education they received: first, change the tools rather than engage in a war of ideas; second, invent new tools rather than philosophical systems. With the strength of these basics as their foundation, they tried to colonize the world.

In doing so, their system for organizing reality went through three significant transitions.

1. To begin with, they went back to recover something that was from the origins of their history, something with an aura of childhood memory. It was a toy. Or, rather, it was a computer that played a game. A video game. Their founding mythology was based on that video game: they saw in it a genetic trait

which had accompanied them from their earliest moments, and which they wanted to bequeath in all the tools they went on to invent. It wasn't a simple feature to pass on, but they worked long and hard, in an increasingly sophisticated fashion, until they came up with new tools that were somehow FUN, but——even more importantly—WORKED WITH THE SAME LOGIC AS VIDEO GAMES. This logic relied on: rapid sequences of actions and reactions, a learning curve based on repetition rather than abstract instructions, the constant presence of a points-system, minimal physical resistance, and sensory pleasure throughout. It wasn't just the nostalgia of a people who missed their childhood. As these features became more commonplace, it paved the way for the idea that solving problems was an action that required simplification, clarification, or summarization. The prerequisite was that the complexity of any problem should be predigested into elementary particles and placed in that form on the surface of the game table, preferably in a manner that made them easy—even fun—to use. Once again, *Space Invaders* was kicking the table-soccer table out from under them.

2. These settings and features were most likely what made the second move, which was completed at unprecedented speed during the same period. Theoretically speaking, this move was practically impossible: dismantling the mental paradigm of the twentieth century and turning it upside down, rejecting profundity as the seat of authenticity, and shifting the core of the world to the surface. The icons on computer and phone screens reminded everyone that the essence of the actions we were engaged in could be mined—down in the illusory depths, where castes of high priests used to be on guard—and brought up to the surface in the form of bright icons floating in broad

daylight. Once you have learned something like this on tools you use hundreds of times a day, you begin to adopt it as a potential strategy for life. It may not be the only one you go for, but it is certainly one of the best on offer. This was another shock, which wiped out centuries of geography and experience, and reconstructed from scratch the art of living by choosing superficiality as its ideal habitat.

3. It is important to add that none of this would have been possible if human beings hadn't continued to blindly believe in the effectiveness of the *human-keyboard-screen* posture. Not only did they believe in it, they went on to perfect it on their smartphones, tablets, e-books, and video-game consoles, relentlessly seeking a result that could well have seemed prophetic at the time. Their ultimate aim, in fact, was to minimize the distance between the three elements of the *human-keyboard-screen* posture and meld them into a single action: POSTURE ZERO-ZERO, the absolute minimum of the model they had envisaged. With smartphones, they attained remarkable results, achieving a kind of utopia that had been in their sights since the dawn of the digital insurrection. That is, the vision that computers may one day become organic products rather than machines, actual extensions of human beings rather than artificial objects. In the book by Stewart Brand, which Steve Jobs worshipped, computers were seen in this visionary light: they were listed alongside tips for growing tomatoes or advice on having a natural home birth. The seed of this vision had been sown, and a few years later, the shoots sprung up in the form of tools like the iPhone. With its touch screen, the *human-keyboard-screen* posture had effectively been reduced to a kind of POSTURE ZERO to which all other postures owed their origins. Once that extreme reduction had

> taken place, once the almost mystic simplification had been realized, circulating between this world and the otherworld effectively became natural and organic. Moreover, the system of reality with twin pumps effectively became an almost natural landscape, a context where there is no background noise, a game table that seemed to have been there forever. This achievement was their third move.

It is no surprise then that this lucid occupation strategy shocked the lumbering old twentieth-century culture out of its listless slumber when it sensed that something big was about to happen. There was a reaction, which we recorded as the first real war of resistance to the digital insurrection and which dated back to the years bridging the two millennia. Generally speaking, the resistance fighters didn't perceive the whole process, they just saw the final results. They could make out enemy footprints, but they never actually glimpsed the enemy. This made their guerrilla warfare more complicated, if not impossible. However, there was probably another reason they were defeated. This was the fact that their most potent weapon—the accusation that the soul of the world was being lost; that there had been a desertification of meaning, true experience, and intensity—was ultimately futile. In their view, the blend of superficiality, aversion to educators, veneration for shortcuts, adoration of video games, and skepticism toward any theory was an evil omen that presaged an intellectual—perhaps even moral—apocalypse. However, it later turned out—though it was hard to appreciate this immediately—that the digital insurrection was equally able to generate experience, intensity, meaning, and vibrations. The insurgents did it in their own way, starting with their capacity for bringing the essences of the world out into the open. Once the essences were on the surface of things, they were able to elaborate them, launch them onto networks, or just make them move about. This turned out to be a perfectly sound way to make the world vibrate, even though the skills were new, untested, and perhaps still incomprehensible to many of the inhabitants. The potential for post-experience was thus revealed, making the twentieth-century model—the only model that had promised access to the meaning of things, though at a very high price—ultimately obsolete.

We can now draw an area on our mappa mundi which is still uncertain and undefined, but real, and which did not exist previously. We will call it post-experience. It complements the other continents we have explored and discovered because they share a prevalent genetic trait deriving from video games. They also share the upturning of the twentieth-century pyramid, the reinvention of superficiality, reverse thinking, and the arrival of the posture zero—the mother of all moves. As you can see, together they form a solid, sound, balanced geographic figure, right? If we superimpose it onto the first mappa mundi, the features registered by the first mountains to break out from under the earth's surface—veneration for movement; direct contact with reality; opening up to the otherworld; the discovery of the first posture, which would later become the posture zero—what we see is a world that is still perhaps a little imprecise in detail with only approximate measurements of distances, but nevertheless a cogent world with a genealogy that can be traced right back to its origins: a stable world with a recognizable form.

Long ago, when the early sketches of just-discovered lands reached cartographers in a form that was as complete, organized, and beautiful as this one, they thought the new lands deserved a name. This was a way to certify, if you like, that these lands had been saved from dark ignorance and become part of our world's fallible knowledge. It was a noble gesture, giving a name.

The Game.

We now know that it was precisely during the Era of Colonization that most human beings migrated and chose this new land as their habitat. They had no maps to guide them; they just moved there, following the tools that showed them the way. A movement that started out as a nomadic insurgency began to settle, seeking out the most fertile soil for their unique construction methods. The great institutions of the previous civilizations were not altogether razed to the ground; they were left to function in their oblivious and inexorable way. The settlers didn't bother to correct them; they simply bypassed them completely, building new living quarters, new cities. They went on to construct towers and fortresses, new defense lines, a new system of governance. People were called upon to manage these, and the players best suited to the Game evolved into leaders. Over time, the first generation of natives was born, human beings who had not migrated there. The first children of the Game. In these digital natives, the Game slowly starts to shed its revolutionary roots. It rejects the legacies of the twentieth century and turns into a game of skills with its own set of justifications for existing—no longer in opposition to an enemy. It is no longer a move against someone, but rather a move toward something. In this sense, it loses some of its ideals, but it gains in efficiency, reliability, and stability. It starts rolling with remarkable skill, and as it rolls, it forgets the obvious fact that not all human beings like that way of being in the world. The inhabitants are too visionary or too blind to notice that post-experience is hard work, destabilizing, and tiring. There follows a widespread sense of dismay that nobody predicted and that soon presents its case.

Almost as soon as it was founded, the Game was already generating discontent.

This is a fascinating moment, because we have reached the gateway to the era that is ours today. It feels like no time ago that we were struggling to grasp the difference between the web and the internet. Just imagine! Or what digital *actually* meant. And yet here we are, in front of the last set of ruins we have to explore. They are spectacular, and we know them very well. They are the houses we live in.

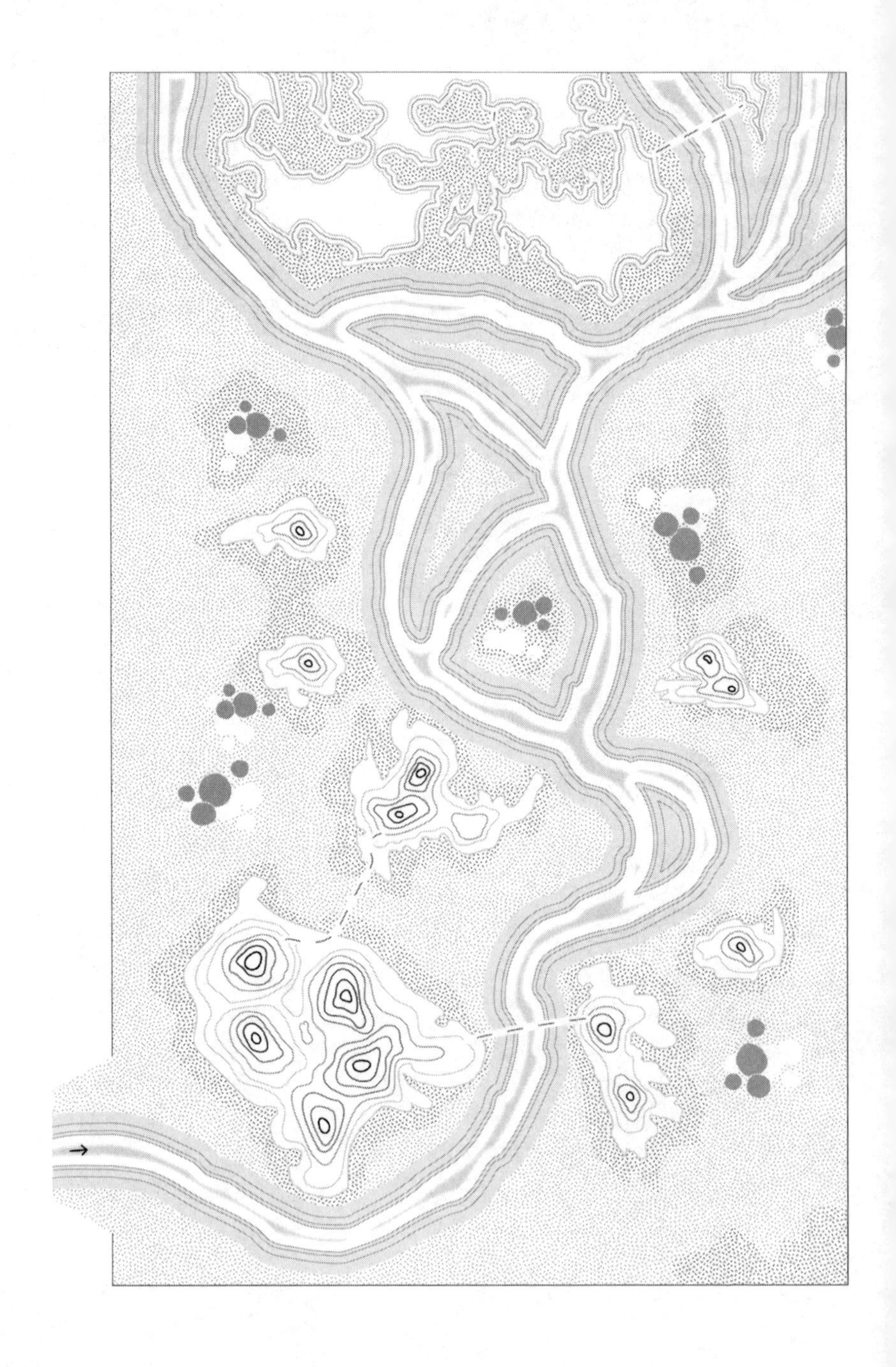

2008–16.
FROM APPS TO ALPHAGO

THE GAME

The world we live in

2008

• In September, Lehman Brothers crashed. We were soon to find out that this was merely the beginning of an economic crisis that would persist for years. Again, the big players in the Game were not particularly affected. Their sales dipped a little but then took off again, unscathed. If anything, they rode the wave of inertia created by a relentless slump in consumer consumption. An almost mythical aura of invincibility was born, as was the instinct among many human beings to see these dominant players as dangerous, fear their omnipotence, and wish for their ruin.

• The Swedish startup that invented Spotify unexpectedly featured on the list of monotonously all-Californian companies. The music streaming platform soon became a model for many other inventions. Spotify's founder was Daniel Ek. He was twenty-five years old and had started making money on the web at thirteen (I swear). When he dreamed up Spotify, the situation was the following: if you knew what you were doing, you could already download any music you wanted in the otherworld without paying a penny. It was called "piracy," and it was illegal, but as you might imagine, chasing pirates in the otherworld is no easy task. In the meantime, musicians and record labels saw their profits dwindle to almost nothing. Ek realized there was only one way out: to do the same thing that pirates were doing, only much better, and to add a little price tag to the activity. His intuition was that if you had all the music in the world at your fingertips for a few dollars a month, you would no longer need to bust your ass downloading music files that you didn't really know how to organize on your computer. The idea didn't come completely out of the blue. Seven years before, iTunes was already out there, offering more or less the same thing. Except that it cost more, it was much less fun, it was stuck in the Apple world, and ultimately, it was incredibly clunky to use. Ek had in mind a basic video game that was quick as lightning—almost magic—and very cool. He managed to build a platform that made this vision come true and ended up beating Apple at its own game. In 2011, Spotify hit the American market and has never looked back. Today, the premium streaming service costs $9.99 a month and gives you free access to over thirty billion pieces of music. The consumption habits it relies on, and the business model it has established, can be seen up and down the vertebrae of the Game (Netflix is no different). We can criticize these platforms as much as we want, but that is not the point. The point is that Spotify and Netflix are typical products of the Game's maturity. They are spectacular interpretations that could only have been achieved in the presence of the logical and technological premises of the previous decade. In other words, they didn't invent anything new, they simply completed other people's unfinished work and added a brilliant twist to the original idea.

• On July 10, Apple incorporated an online store on their devices for the first time, selling what were called *applications*. Nowadays, the world over, they are called "apps." There were about five hundred of them on the crowded online shelves. Very few of them cost more than ten dollars, and a quarter of them were given away for free. You downloaded them onto your device using your internet connection. Easy peasy. Now, would you like to know how long it took for the inhabitants of Planet Earth to download ten million apps?

Four days.

Apple was thrilled to post a press release stating that the app situation had blown out of all proportion. They had no idea the genie had been let out of the bottle. The last time I checked, back in 2017, people were downloading 197 billion apps per year. The figure is probably much higher now. The Apple Store is still going strong with more than two million apps on offer, but it is by no means the biggest store in operation. Google Play, for example, has a catalog of more than three million.

The only thing we need to do now is understand what an app *really* is.

It's not easy for us mortals.

Just to give you an idea, apps have been around for a good while in the Apple Store. The system we used to send emails from the start was basically an app. So was Microsoft Word. We called them *programs* (or software, if we wanted to cheat a little). They were long sequences of commands that told the computer exactly what to do in order to complete a specific task, such as sending an email or writing a paper. There are three reasons why we stopped calling them programs and started using the word app, which is much more pop. First, app was easier to pronounce. Second, we started to invent programs that were much cooler than MS Word and were definitely the native products of the Game. There were thousands of video games, of course, but also programs that reminded you when to go to the bathroom or identified the music playing at the supermarket or turned your photos into Van Gogh paintings. The third—and most important—reason was that new programs were being developed directly for smartphones rather than for computers, which meant they could be carried around in your pocket. These new programs covered a whole series of needs and desires that could no longer wait until you got home to your personal computer to be satisfied; they had to travel around with you. You didn't even need to switch your computer on: they were there, on your cell phone, which was in your hand; all you had to do was tap on the icon and—hey presto!—it was done. "App" was the perfect, almost onomatopoeic word for this. Like "ticking" or "bomb."

Once programs became apps, we started using them, loving them, relying on them, and playing with them. Whereas we used to think of them as monsters, now we treated them like pets. One of the consequences in particular of this change is worth noting. Through our apps, we have opened up an infinite number of gates into the otherworld. Where once only the web dared tread, now millions of apps, which don't necessarily have anything to do with the web, venture forth. They often don't have a website address and cannot be reached via the web. They are like closed clubs we only go in and out of to get something we want. These closed clubs are in the otherworld, however, and the sheer quantity of them circulating the planet sends a very important message. That is, there is a lot of traffic over there, in both directions, and it is very fast indeed. This means that preserving the sense of a border between the world and the otherworld has become impossible and almost always meaningless.

This is important. Take note. Once you can no longer distinguish where the border is, you are in the Game.

• Airbnb was born. The startup was a perfectly logical extension of the old idea of cutting out all mediations and gaining direct contact with reality. Do you have an apartment you're not using? Great! Stick it on a noticeboard in the otherworld and rent it. More and more people were doing it, and three young Americans had the idea of creating a website that would match people who wanted to rent their property to people looking for a place to stay.

The name came about when these penniless guys with a small apartment in San Francisco inflated three air mattresses on the floor of their sitting room and started renting them out: Airbedandbreakfast, they called it. A bit too long, right? They abbreviated it to Airbnb.

• On November 4, Barack Obama won the US election, making him the first African American president in history. The reason he is on this list is because he was also the first politician to use the digital world to aid him in winning an election. It wasn't just one of the many media outlets he relied on, it was the "central nervous system" of his campaign.

The heart of his operations was his personal website, MYBO

[*sic*], and in very little time, he built up a phenomenon that was neither a party, nor a campaign, nor an organization. It was a giant community of people who shared a dream—Obama for President—and were in possession of the tools they needed to meet up, recognize one another, exchange information, and lend a hand. On the website, for example, there were twenty thousand groups: you could choose the group you felt the greatest affinity with (tango dancers, single moms, etc.), and you immediately became a part of this little clutch of people who were like you. If you felt like lending a hand, the site gave you all the contact details for people in your area who might be unsure about voting Obama, and you could phone them or go and pay them a visit. Whatever you wanted

to do. There was another part of the site devoted to fundraising. They didn't ask you for money. It was much more fun than that: you yourself could become a fundraiser for Obama's campaign. You set a goal—say $10,000—and then you started working on your friends and acquaintances. A nice big thermometer measured the progress of your mission. I've already said this: if things didn't look and feel like a video game then, you didn't get anywhere.

There were many people who contributed to inspiring and building this system, but there was one guy in particular, Chris Hughes, who was the brain behind the whole operation at the age of twenty-four. He had been one of the four founders of Facebook, the only intellectual in the group. He hated Silicon Valley and preferred the East Coast, where he graduated in history of French literature (!) and then left Facebook to go and work with Obama. It's also worth noting that the machine room of MYBO was inside Blue State Digital, founded in Silicon Valley with offices in Washington, DC, and Boston. This is significant because it helps us realize that nobody gave a damn about politics in Silicon Valley. It was only when people started going to the office on foot, a few blocks from one of the greatest seats of power in the world, that the idea of applying digital techniques to something as obsolete as politics became at all credible.

2009

- WhatsApp was born, almost by chance. Originally, the idea was to create an app that allowed you to add a little phrase like "leave me alone today" or "I think I'll go to the pool now" after your name when it appeared on other people's contact lists. It was a cute notion, which the two inventors, Brian Acton and Jan Koum—who was born in Kiev and moved to Silicon Valley at sixteen without a penny to his name—soon realized had far greater potential. They had designed an extremely easy, efficient messaging system. Apart from the classic American immigrant story (they are both billionaires now), it's interesting to note that both Koum and Acton worked at Yahoo! and that the idea came to them when Jan got his first iPhone, found the Apple Store, and discovered there was a market for apps. What I'm trying to say is that they were second-generation inventors, as many inventors in those years were. They worked inside a system and developed its potential; they didn't develop systems themselves. It was the Game beginning to produce a new generation—it was going forth and multiplying on its own.

WhatsApp was sold to Facebook in 2014 for nineteen billion dollars.

Today, more than a billion inhabitants on Planet Earth use it regularly.

The inventors chose never to accept advertising on their app.

Since 2016, it has been completely free.

How the heck do they make any money, then?

If you're looking for an easy, consoling answer, here it is: they sell our data.

However, this has not been demonstrated. In fact, it was against the principles of the two inventors and excluded by the Terms of Service you accept when you download the app. There may be some doubt about posted photographs, but your WhatsApp messages are neither read nor commercialized.

So, how on earth do they pay the servers that help all these messages circulate, you may well ask?

As far as I've been able to gather, the most reasonable answer is the following: if you are in a position to make daily contact with a billion people in the world, you'll always find someone to lend you money, and sooner or later, you'll find a way to pay them back.

- Uber was born, greeted with jubilation by taxi drivers across the world. It followed fast on the heels of Airbnb in the same city (San Francisco). It seems that the idea floating about in that city was that if you had anything you weren't using, you may as well give it to somebody else and make a little money on the side. Do you have a car and a couple of hours to spare? Become a taxi driver without really being one, and drive taxi drivers out of business. If you amplify this concept, you get the Sharing Economy, which was one of the trends launched in the years of the Game. It was the result of the instinct to cut out all mediations, but it adds a significant twist: the idea of shared property. It started out by renting an extra room and developed into co-housing, car-sharing, and crowdfunding. The digital insurgency's hippy forebearers would have been pleased as punch, though the castes that controlled consumer services were less happy. Of course, when some people organize and share what they own—without recourse to experts, mediators, high priests, and license-holders—other people are going to lose a great deal. In this sense, the legal (and sometimes physical) skirmishes between Uber and taxi driver unions represent a struggle that took place around the world back then. We don't need to take sides here. On the other hand, it's good to know that none of this would have happened if digital platforms designed specifically to encourage the exchange of goods among ordinary people—and thus trigger revolt—had not been invented.

One more thing: the guys who invented Uber were already incredibly rich when they came up with the idea. They had both made money selling relatively successful startups. One was a peculiar search engine (StumbleUpon), the other, a file sharing platform (Red Swoosh). They were not second generation, but they were on their second round. It was here that it started to be clear that the insurrection was no longer the prophetic vision of a few outsiders; it was becoming a neurosis of a small, recently formed elite.

• On October 4, in Italy, a political fraction called the Five Star Movement (M5S) was launched. It was the first time that the digital insurrection had directly spawned a political group with the declared aim of storming the palaces of power. As we have seen, this move was not entirely congenial to their nature. The digital insurrection's invasion strategy was, if anything, to bypass the main twentieth-century institutions altogether (politics, education, religion, etc.) and to dig underground tunnels around them while developing more and more new tools. Obama's election campaign, in fact, offered digital tools to those who wanted to lend a hand, but the values and principles of the aspiring president resided within the solidly twentieth-century tradition of the Democratic Party. The Five Star Movement was different. In this case, a segment of the population of the Game was getting fed up with a listless, corrupt, and outrageous political class and decided to take things into their own hands.

The digital DNA of the early movement was very strong. There were two people behind it. One was a popular comic actor called Beppe Grillo, who had kept a phenomenally successful blog going for the previous four years (it was classified among the ten most influential websites in the world by authoritative sources). The other was Gianroberto Casaleggio, a computer programmer for Olivetti who later became a consultant for companies seeking to establish themselves on the web without knowing what it was (the companies, that is, not Casaleggio, who knew exactly what he was doing). Grillo did a great job of galvanizing the latent energy that was smoldering in the gloom of a disenchanted country, mortified by the indecisiveness of politicians, and suspicious of the dominant economic powers. Casaleggio, in his turn, knew how to create the tools to channel that energy and give it political form. In 2007, civic lists generated by communities of Grillo's blog followers were presented at regional elections. That same year, a mass demonstration with the sophisticated title "Fuck Off Day" filled squares across Italy, giving the movement the visibility that they had previously lacked and earning the Grillo followers a name: *Grillini* (or "little crickets"). Two years later, the movement was launched. The five stars in the name have nothing to do with hotel classifications. They are a reference to the movement's five priorities: water, the environment, connectivity, development, and transport. These were Grillo's hobby horses which, translated into policies, meant: a citizen's wage for everyone, sustainable degrowth, access to the internet as a basic right, environmental protection, and a lifestyle with a zero carbon footprint. Casaleggio's contribution was the idea of digital democracy—a form of direct democracy—which, following the axioms of the digital insurrection, cut out all possible mediations and encouraged people to intervene directly in the political arena by means of their digital devices. It may have sounded utopian at the time but, if you think about it, it was a slightly ambitious but fairly logical upgrade of what was happening anyway. In a world with Uber and Airbnb, where the biggest encyclopedia in the world is written by everyone, where anyone can respond to the pope's tweets, and where news reaches you through Facebook, why shouldn't people vote with a click or become political representatives themselves without belonging to any elite? Having said this, the Game is not made up of ideas, but of tools; therefore, imagining a digital democracy means nothing without creating a platform that makes it possible. Casaleggio did precisely that. He built Rousseau. It doesn't have a great reputation, it must be said. It was evaluated by Italy's Data Protection Authority as "completely lacking the requisites of IT security that should characterize an electronic voting system." The inappropriateness of the fact that the person who invented the platform also owned it has also been commented on (after Gianroberto Casaleggio died in 2017, ownership passed on to his son). It's undeniably true that in a world where the web had been bequeathed to everyone for free, it's curious that someone should own the keys to the tool that is supposed to make direct democracy work. In any case, the platform Rousseau exists and still regulates the direct democracy of the Five Star Movement. As far as I am aware, there is nothing similar anywhere else in the world.

The Five Star Movement ran for regional elections for the first time in 2010 with results oscillating between 1 and 7 percent of the electorate. Eight years later, it was the most voted entity in the 2018 national elections, even though it wasn't a party, with 32 percent. While writing this book, the movement was in national government. Unfortunately, 32 percent was not enough to govern on its own, so they had to find allies. The fact that the movement started out with the aim of turning Italy upside down, giving every Italian the password to gain access to power, meant that there weren't too many parties willing to join a coalition with them. The result is that they are finding it hard to move forward. It's as if they were trying to connect an iPad to a tractor with a USB pen. They need to be patient. It's hard to tell how things will end. If anything happens, I'll let you know. You might be really interested and want to know how the heck we Italians have gotten out of the situation.

2010

• Instagram, a much cooler social network site than Facebook, was born (only to be bought up by Facebook two years later). Its inventor, Kevin Systrom, graduated from Stanford, worked at Google, and had already invented an unsuccessful app for iPhones called Burbn. He was a perfect specimen of a second-generation innovator: just for a change, he was a white, American, male engineer.

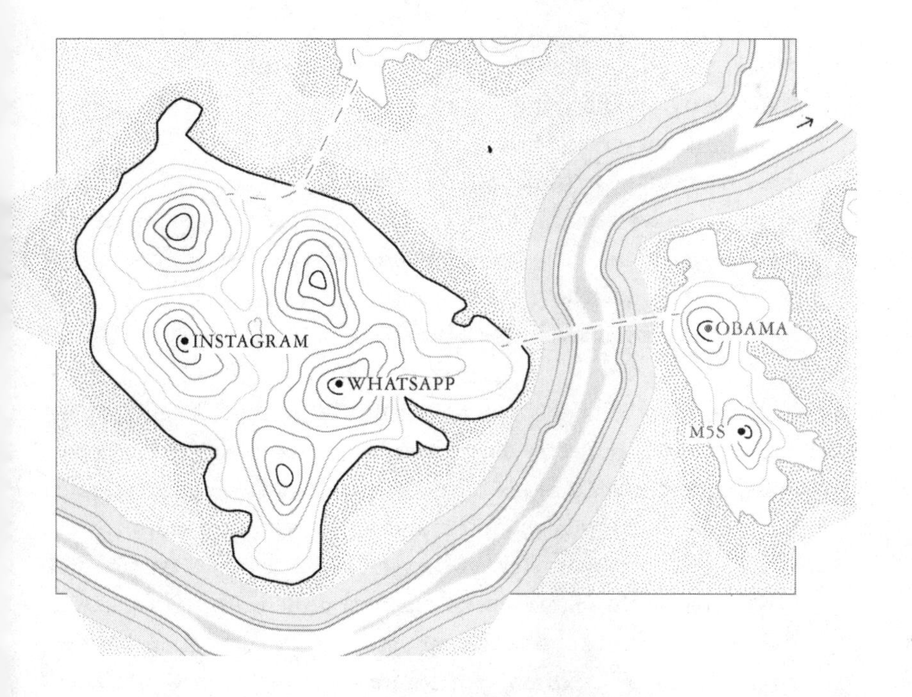

2011

- Apple launched iCloud, a system for storing the contents of your computer (or iPhone) in a place that was not your computer. You sent them off onto a cloud: when you wanted to access them, you brought them back, and then, when you no longer needed them, you sent them back onto the cloud again. Aside from the technological angle (it's not very nice to point this out, but your files, your contacts, your love letters, and your photos in boxer shorts were not sitting on a cloud; they were stockpiled, in the millions, on other computers in the most absurd places), the symbolic value was remarkable. We wanted to dematerialize reality, and you could say that, with iCloud, we succeeded. A vast part of our life now weighed nothing, lived nowhere specific, and followed us around without occupying space or time. Astonishing. In practical terms, the advantage was that, if your smartphone fell into the toilet, you could flush it down and still have your contact list intact. The disadvantage was that strange sensation of leaving everything you own hostage to someone you don't know. Like when you want to go and have a swim in the sea but you're not too happy about leaving your backpack—with everything you own in it—on the beach and a sunbather next to you says, "Don't worry, I'll keep an eye on it for you." Well, it's a bit like that.

• Three students at Stanford invented Snapchat. To all appearances, it was a perfectly normal messaging app: elementary, easy, and direct. Perfect for kids. They soon added an awesome variable, however: the messages, photos, and videos you sent on Snapchat vanished into thin air after twenty-four hours. It was irresistible! Given that in the Game, one of the most difficult things to do was to hide, vanish, change your mind, atone, erase, etc., Snapchat was an instant success. Currently, there are approximately two hundred million daily users.

• 2011 was also the year that set an interesting, in some ways crucial, record. I have no idea how we know this, but this was the year that apps were more used than the web. I'll be more precise: from that year on, we were more likely to tap on the icon of an app than on the icon of the browser that takes us onto the web. Professor Berners-Lee can't have been too pleased. It's not that he wouldn't have approved of apps overtaking the web; it's simply that he had dreamed of the otherworld as an open space that belonged to no one, where human beings were able to exchange whatever they owned. Apps do not correspond to that description. They belong to their owners, and they are not an open space. They are like warehouses: giant, closed warehouses. People go in to get a certain service, and then they go back into their dens. The difference is clear. If you like,

it's a symptom—among others—of a kind of cheerful degeneration of the Game; a slow erosion of its original utopian force. However, readers, beware: the chances that this evaluation is the product of a moralistic, consolatory, twentieth-century analysis are very high.

2012

• After music, photos, and videos, TV also officially became digital. Analogical TVs went out of the window. Nowadays, the only country with no digital TV signal is North Korea.

• Tinder was born. This app finally freed the genie: a lame desire that nearly everyone had felt, at one time or another, was at long last out in the open. You could choose a partner you had never met before from a catalog and go out to dinner with them (or go to bed with them, for that matter). It was not the first time that dating sites went online, of course. Tinder was the first to understand that choosing a partner from a catalog while you were lying on the sofa with the TV on was, for the vast majority of human beings, more fun than actually going out to dinner (which was expensive and, to make matters worse, forced you to wash and get changed first) or actually going to bed with someone (I'll desist from listing the innumerable issues at stake there). The inventors of the app were careful to make the catalog of potential partners as similar as possible to a basic, easily played video game with an erotic twist. It was like a game of solitaire with a pack of cards. The fact that you may end up necking with the Queen of Spades by the end of the evening contributed to the app's success.

2016

• Between March 9–15, software developed by Google (yes, them again) challenged the world Go champion—Lee Sedol, a South Korean of thirty-two—to a best-of-five match broadcast on YouTube with a million-dollar stake. AlphaGo won four matches to one.

A machine beat the top player in the world.

I know you're all thinking about Deep Blue's victory against Kasparov (it was 1996, and the software had been developed by IBM), but I invite you to think about the difference: Go is an immensely more complex game than chess. For example, the first move in chess has twenty possible variations, while there are 361 in Go. If you manage to survive the first move in chess, the second has four hundred potential variants, while in Go, the choice of your second move must be made from 130,000 different possibilities. Good luck to you!

This is just to say that designing a machine that is capable of making such complex choices takes a lot of work. Training AlphaGo, for example, required the software to memorize the moves played by (advanced) human beings in thirty million matches. Up to that point, however, it was just a matter of muscle power; the machine's ability to make advanced calculations was no particular surprise. The interesting part was when the programmers started working

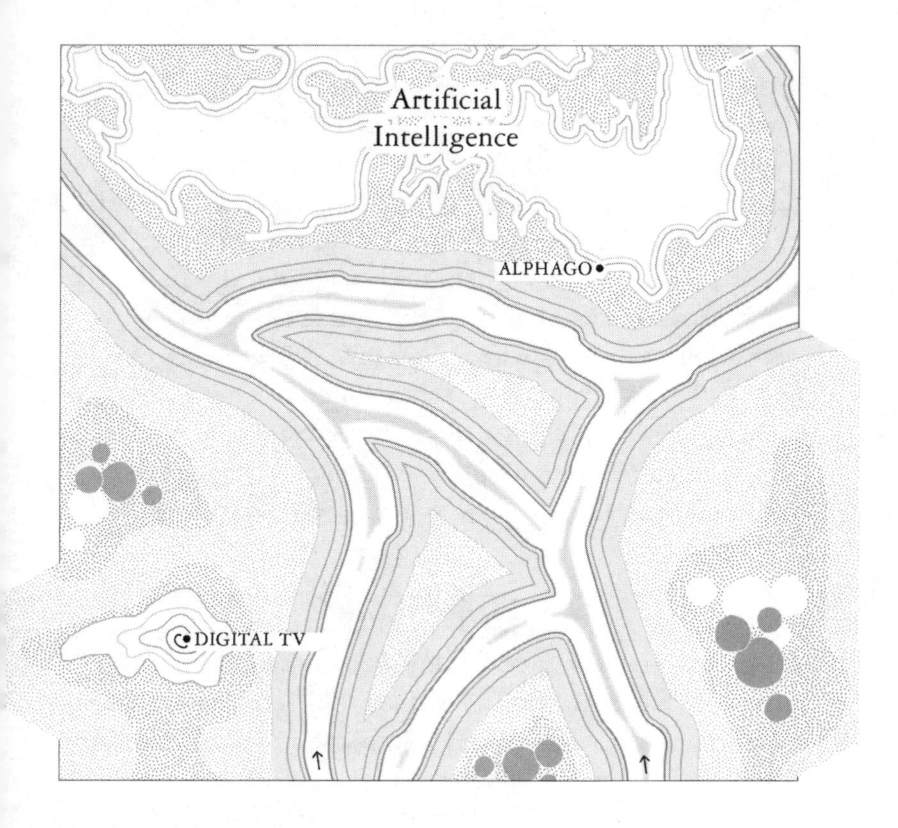

on the software's deep neural networks (there's no point trying to understand what that means), which was when they achieved a result that was truly extraordinary. AlphaGo had learned all its moves from human beings, but it started inventing moves of its own accord and applying strategies that human players had never imagined. That was why the machine won.

There's a name for what AlphaGo embodied, developed, and brought to the attention of the world: ARTIFICIAL INTELLIGENCE. I'm not even going to try to explain what this means exactly; I'll think about it in ten years' time, when I get around to writing the third volume of the Barbarian saga. The fact is, I needed to get the name in here because it marked an important watershed, a new technological transition, perhaps even a new mental perspective. With artificial intelligence, the chapters of the Game we have been reading give way to a new era which, frankly, is hard to make predictions about. If you are envisaging robots pissing on your head, you're on the wrong track (that kind of artificial intelligence is still stuck in the previous era). The rest is a wide open horizon yet to be explored. We shall see.

For now, let's take home the pleasurable satisfaction of having drawn a line that starts with a game involving little Martians and ends with a game played by South Koreans. Linking these two points, I can see the elegant range of mountains, representing the backbone of the civilization I live in, unfold before my eyes. Is there any reasonable hope that you can see it too?

FINAL SCREENSHOT

We are on the last crest of the range. The sensation is of a civilization that has laid its foundations, established its cornerstones, and is now tightening the nuts and bolts of the structures it has built. There was work to be done, and it has been done.

It will probably be remembered as the era of apps. The transformation from baffling, clumsy old ogres (bold software) into little lap dogs that cost almost nothing (apps) completed a series of moves that were initiated years before:

- The traffic between this world and the otherworld melted, dissolving with it the psychological border that in the previous era kept the two regions apart;
- The system of reality with dual pumps that was first experimented with by the web was brought to completion;
- A whole range of mediations and mediators were definitively cut out;
- A habit was formed whereby solving issues was required to be fun, with the result that daily problems were melted into a sea of little video games;
- The sensation of being part of an augmented humanity became generalized;
- Access to post-experience became easier;
- Absolute mobility became the rule, giving privileges to smartphones over computers, making the *human-keyboard-screen* posture increasingly light;
- The distance between human beings and machines was reduced to a minimum to the extent that devices became perceived as organic products, "natural" extensions of the body and mind.

It's a lot of stuff, right? There are lots of impulses that combine together to create a way of being in the world, lots of intuitions that fit together like a puzzle, creating a single design. Forty or so years of rebellion channeled into one Promised Land: the Game.

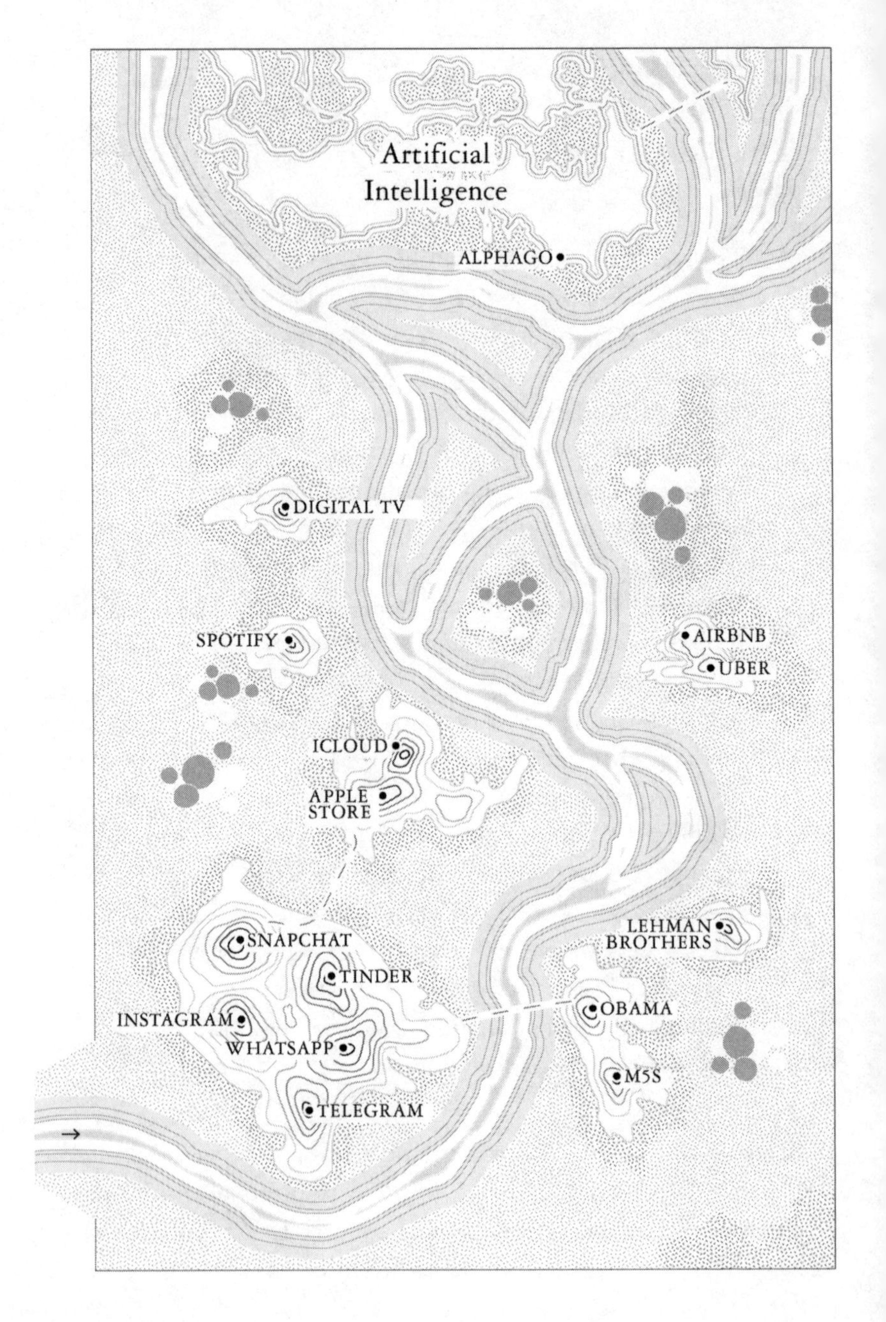
Artificial
Intelligence
ALPHAGO
DIGITAL TV
SPOTIFY
AIRBNB
UBER
ICLOUD
APPLE
STORE
SNAPCHAT
TINDER
INSTAGRAM
WHATSAPP
TELEGRAM
LEHMAN
BROTHERS
OBAMA
M5S

The homeland that was so hard fought for was complete. It left behind the nightmare of the twentieth century and gathered together its forces in order to prepare for the next thrilling colonization. The surprising advances in artificial intelligence pointed them in the right direction.

To sum it up, one could easily say it was an era of triumph.

And yet, if we dig further down into the ruins and extract the archeological remains, we can see a different, slightly more complicated story when we dust the artifacts off. While the settlers may have appeared happy with the work they had completed, the contradictory underground traces tell another, more destructive story. There are signs of demolitions, struggles, and earthquakes here and there. The overall picture is harder to read, undermining our certainty that the Game was a complete system. New questions emerge, some of them with no answers.

As we shall see in the "Commentaries" that follow.

COMMENTARIES ON THE ERA OF THE GAME

MASS INDIVIDUALISM

The concept of *augmented humanity* first came up many pages back. We were examining the early years of the digital insurrection and used the expression to describe the buzz every user felt in the early days of the web: the thrill at being able to move laterally, travel anywhere, nose around in other people's desk drawers; the new frontier of the otherworld that became accessible to all; the speed; the freedom.

This was in the early nineties. Thinking back to those days makes me smile. Now that we all know what happened in the twenty-five years that followed, it seems absurd that we used the expression *augmented humanity* for so little. We couldn't even send an email on a smartphone, yet we were already calling it *augmented humanity*. We were already debating whether these things were bad for our mental health, for society, for the world!

How cute!

Now that we're in the Game, we can see things more clearly. It's our duty to understand what has happened.

We know for sure that a generic penchant for developing individual potential became more manifest in the Era of the Game, where an immense network of tools had effectively multiplied people's opportunities. We don't really need to go into great detail here but, if you like, take four examples of actions—traveling, playing, getting information, and having relationships, say—and compare the tools we use today when we do these things with the tools we had twenty years ago. There is an abyss between the two eras; no kidding. Humanity has *truly* been augmented.

Another thing we know for sure is that a spectacular upgrade of this kind did not take place only in a restricted elite; it involved practically all human beings on this planet. OK, you don't agree with the fact that it involves *all* human beings; you're right. However, if you take into account one billion WhatsApp users, two billion Facebook accounts, and five million rental properties listed on Airbnb, you must accept that we can't say it only involves the elite, the rich, the cool, or the West. We are talking about something that concerns a striking number of people. We are used to it now, but the idea of opening up access to all that potential, across a social base as wide as this, was a colossal enterprise. The distribution of *potential* was also the distribution of *power*. Thirty years ago, only the most ardent hackers, who had been cultivated and nurtured in the fertile oil of the Californian counterculture, could have imagined it. We now know they weren't "totally off their heads." Incredibly, their idea of using computers to break up centuries-old privileges and redistribute power among all human beings made sense. I myself wouldn't have bet a dollar on it, I swear. And yet.

The third thing we know for sure is that this redistribution of power logically overlapped with another feature which was present right from the beginning of the digital insurrection. That is: the desire to cut out mediations, have direct contact with reality, and disarm the elite. The two forces worked hand in hand for years: the more individuals received redistributed shares of power, privilege, potential, and freedom, the more they used these shares to liberate themselves from useless mediators. Try multiplying this dynamic by a billion and you'll see what we are talking about. It was a spectacular geological reversal. An earthquake, if you like.

Naturally, the world came out looking very different. We are now in a position to see in what way it changed. It's vital not to pause on the details or get distracted by bizarre features that emerged, such as being able to organize a trip without a travel agent or reader forums becoming more influential than literary critics. Who cares? These features are not the point. We need to distance ourselves a little, look down from above, maybe. We can thus see the heart of the matter, the precise point where the earthquake redesigned the shape of the world. It's a point we know well.

It is the point where self-awareness takes shape.

What happened in the Game is that billions of individual human egos were being fed a daily diet of super vitamins, in part generated by the tools that made them far more effective, and in part developed by the constant patricides they committed while liberating themselves from the elite. A new self-awareness came out onto the surface of billions of people's consciousnesses. People who were not used to seeing themselves that way, who were not *destined* to imagine themselves in this world, if you see what I mean. In a certain sense, they discovered themselves looking at reality from the front row seats in the stalls they had simply never aspired to or from theater boxes they had always thought belonged to others by divine right. Before, even if they screamed and shouted, you could only hear them from the stairs of your building; now, any whisper can be heard in Australia. All of a sudden, people realized that they could THINK DIRECTLY; they could have opinions, without waiting for the elite to express them first and, only then, share them. You yourself could produce opinions, shape them, express them, and then inject them into the circulatory system of the world, where they might potentially reach millions of people. Only a hundred years ago, no more than a couple thousand people would have had that kind of opportunity in the whole of Europe.

So, we have a hypertrophic ego. Rather, a reconstructed ego, because one of the things the digital insurrection managed to redistribute to the people was an enlarged ego. This privilege was previously reserved for the elite, although they themselves never thought their egos were puffed up; they simply thought they were the most harmonious expression of their brilliance. Let's call it a highly skilled restoration of individual egos. One of the most advanced features of digital tools, in addition to restoring our egos, was creating a protected environment for them: a soft terrain where they could grow more robust without risking too much. All social media, as well as simple messaging systems and massive containers such as YouTube, were studied with the precise intent of allowing you to come out into the open but *not too far out into the open*. They allow you to express yourself, slightly ambitiously or even aggressively, without ever leaving your *comfort zone*. The context is ideal: some people use these tools to insult a government minister by calling her a slut, others to share their first three songs and become famous. In any case, they are an opportunity that thousands of individuals have decided to exploit. At this point, we need to understand this one important fact: WE ARE TALKING ABOUT INDIVIDUALS, millions of individual human beings. Believe me, this is unprecedented. What I mean is that, in the twentieth century, for example, an individual could feel part of an "augmented humanity," but only ever in a context of a collective ritual or if they belonged to a particular community. You could experience moments of great intensity, of self-expression, even of grandeur, but they were mostly moments that you experienced as part of a community such as your nation, your church, your party, or the smaller community of your family. Augmented humanity, in that context, was a collective upgrade, not an individual one. Individuals, on their own, did not go very far back then.

In the last thirty years, however, something has changed. Something enormous. The Game allows only individual players. It was designed for individual players; it develops the skills of individual players; it gives points to individual players. Trump and the pope send individual tweets in the full knowledge that the inhabitants of the Game are used to individual profiles and to playing one on one. The Game has thus become a fantastic incubator for an unprecedented mass individualism. We don't know how to deal with it, and we are basically unprepared to tackle it. I'm not sure whether anything of the kind has ever happened before on this planet. The only precedent I can think of was, perhaps, democracy in fifth-century-AD Athens, which was a regime of mass individualism of a kind. However, "mass" at that time meant 15 percent of the population of Athens. It was enough to create terrible havoc (as well as exquisite beauty), but it was still only 15 percent. In Italy, to go back to my own backyard, one out of every two people has a Facebook profile.

We thus find ourselves living in an unexplored territory where nobody has ever played a match to the bitter end. We keep having to play absurd games that we find hard even to grasp. When individualism becomes a mass phenomenon, in fact, the first thing that goes out of the window is the concept of "mass." What I mean is that there are no longer social groups that move around as groups where something above individual level binds you together in a sense of belonging, such as being Catholic, American, communist, or lovers of rock, for example. In the past, people were more like big animals, moving in herds in an almost impersonal fashion, protected by the taming effect of belonging to a community, and guided by the elite that shepherded them with a strong hand. In the Game, this herdlike movement is rare because mass individualism generates millions of micro-movements and takes the crook out of the shepherd's hand. Nowadays, following the precept that movement is of the greatest value, big globs of consensus coagulate in a moment and then thin out just as quickly. This is because they are not geological formations that have sedimented over time; they are rapid gatherings of individuals, destined to cluster in a different way with the next move. The result: once mass individualism arrived in the Game, the concept of "mass" changed. If a mass ever formed, it was only for sporadic episodes in specific games.

Another paradox that fascinates me is the depressing phenomenon of individualism without identity. That is, human beings who know how to formulate opinions without really having any, who can hand out authoritative judgments without having any competence, who can make crucial decisions about their lives without having any real knowledge of them. It is as if technical capacity had bypassed the actual substance of things; as if digital tools had added powerful motors to the bodies of machines that were not strong enough to contain them, tolerate them, control them, or even really use them. This phenomenon is not entirely new, because in the past we often generated mental systems without having the immediate capacity to sustain them. Take the Enlightenment, for example, which produced a call for liberty that the movement was not able to manage at the time. Similarly, the Romantic movement made sensibility so readily available that most humans could not *physically* tolerate it. It is a familiar syndrome that does not solve the problem of living in the Game, where a good half of the players wants to climb up and perform on stage when they are supposed to be in the stalls watching the show. There's a pushy crowd out there behind the scenes, waiting for their moment in the spotlight.

I don't want to go on too long, but there is one more paradox I'd like to briefly note here, because I think it's important. Individualism is always, by definition, a posture *against* something; it is a seed of rebellion, a claim that you may bring about an anomaly, a refusal to walk with the herd. You are walking in the other direction. And yet, when millions of people start walking in the other direction, where is the path leading?

Did the fathers of the digital insurrection foresee these paradoxes? I don't believe they did. Were they impossible to envisage? Perhaps they were, but a little clearheadedness would have made them easier to predict. Can one live with paradoxes like this? I think we will be forced to admit that we can, since we are living with them today. At the same time, they represent cracks in the surface, like pieces of mountain that unexpectedly fall off the rockface of the Game, reducing its strength, logic, and beauty. These cracks create disorientation and dismay. I would advise against forgetting, however, that they are the result of a move that hoped to liberate the world by creating a clean break with the past. The least that could happen after redistributing power was a less sharply focused, less harmonious, and perhaps less substantial landscape. Thus, if I return to my excavations and observe the ruins of that civilization, I feel as though I can see signs of an incomplete achievement in the era of the apparent triumph of the Game. Having given dignity back to many people and awareness to the majority, these human beings spent the first few years of the Game attempting to regain an equilibrium, to reach maturity, and to bring about a new form of elegance. It seems they were still lacking the ability to be themselves. No tool could give them that, after all.

A NEW ELITE

Talking of paradoxes and curious occurrences, another event is clearly revealed by these archeological remains. That is, after years trying to free themselves of the elite, in order to create a system that was able to sustain itself on the basis of mass individualism, what happened, of course, was that the Game ended up producing its own new, completely different elite, which was nevertheless an elite. A recently revealed fossil tells the story very well. Mark Zuckerberg's Senate hearing in April 2018.

As you will no doubt remember, it had just come to light that a British company, Cambridge Analytics, had used personal data mined from millions of Facebook profiles to influence the 2016 US presidential elections. Caught by surprise, the lumbering old politicians in Washington, DC, who had always let Zuckerberg do whatever he wanted, suddenly woke up from their post-prandial torpor and called the kid in to account for his actions. If you check out the video, the scene is sublime: a row of powerful old men sitting comfortably in their leather armchairs, placed in an almost perfect semi-circle with a second row behind them of what may be their assistants. They are all leaning forward, looking down at the center of the ring where the kid is sitting stiffly, in total solitude, with the sole comfort of a glass of water by his side, as if he were a prisoner under trial. Zuckerberg is wearing a suit and tie, which is significant because he didn't apply the dress code of his company on this occasion, opting, rather, for that of the old codgers looking down at him. He accepted, that is, to play by their rules. Every time he opens his mouth, it's as if he's trying to explain the meaning, origins, and mechanisms of something they know next to nothing about, for the benefit of a bunch of wizened old children. They ask him some surreal questions, and he visibly finds it hard not to laugh. If you reversed the situation, it would be as if he were asking a US senator something like, "Did you become a senator for your own financial gain or because you wanted to help the US?" Or, "Since you were elected, are your electors better off than they were before?" Questions of this kind. The whole thing is a charade. Zuckerberg manages not to laugh; if anything, he looks incredibly tense, playing the part of a high school student who's been called into the principal's office. He's in the weird position of being put in the corner for bad behavior by people he had never taken into consideration before; he's losing a game he's never played before. It's as if he's raising his hands in surrender in front of a firing squad of a dozen men who are raising their rifles to shoot without realizing that their ammunition had been taken away years ago. The dramatic irony is fantastic. Shakespeare was an amateur in comparison.

The old elite and the new in a face-off.

The old elite looks bloated, ancient, irremediably full of themselves, and still powerful. The new elite looks vaguely artificial, cold, almost impersonal, stuck-up, but a little put out by the situation.

It's impossible to say who won. It's like trying to decide whether an eagle is stronger than a leopard. (These are the kinds of questions little kids ask, like whether Spiderman is stronger than Jesus.) The two worlds simply have nothing to do with each other. They have no point of contact on a spatial level. Take one example. It is just one, but it is significant: how they relate to ideology. The senators have an ideological framework, while Zuckerberg does not. The senators would like reality to function according to a few ideal principles, while Zuckerberg only cares about making reality function. The senators get bogged down by the typical American dilemma—how to regulate business without challenging the ideological talisman of neo-liberalism—while Zuckerberg just wants people to hook up with one another. When the senators ask him, in dismay, whether it might not be a good idea to establish some boundaries, as those wimpy Europeans had done, he says that it might well be a good idea. He doesn't give a toss about American liberalism. He wants people to hook up with one another. He seems almost sorry that this desire creates a little trouble here and there. He assures the senators that his technical staff will look into the matter. He doesn't expect governments to be able to help much, but if they have any helpful suggestions, he seems to be implying, he's happy to listen. Period. This is the secular approach—at times scarily beyond repair—of the fathers of the Game.

Zuckerberg's Senate hearing sums up the situation perfectly: it visibly demonstrates the chasm that lies between the two models of power, the twentieth-century one and that of the Game, facing off. It helps us realize that a paradigm shift has taken place, creating a new elite. It would be simplistic to believe that the "new elite of the Game" was made up only of people like Zuckerberg, a handful of billionaires who invented the successful tools that turned our world upside down. If anything, these oligarchs are irrelevant, because the strength of a system never lies in the super rich at the top; it lies, rather, in its capacity to create an extensive, pervasive elite that works at every longitude and latitude to disseminate a certain way of being in the world. In this sense, if you really want to know who the new elite is, look a little lower down the social scale and you will find it. They are humans who are not too difficult to recognize because THEY ARE THE ONES WHO ARE CAPABLE OF POST-EXPERIENCE.

Remember? The intelligent version of multitasking? That way of using superficiality as a terrain for meaning? That technique of dancing on the tips of icebergs? Are you with me still?

OK. The new elites can be easily recognized by the fact that they are capable of post-experience. They move around easily in the Game, using superficiality to propel themselves; they find strength in the impermanent structures created by their movement. They have a capacity for creating chemical reactions in materials that exist throughout the Game in order to forge new materials with which to build new homes for meaning. They use devices organically, biologically, as if they were prostheses. There are no longer any demarcation lines between this world and the otherworld for these new elites; they move between the two like amphibians perfectly adapted to a system of reality with dual pumps. Their mental moves are incredibly fast, and they hardly ever grasp things that are not in motion; they literally don't see them. They do not suffer the destabilizing features of post-experience because they have little or no experience of stability and because they appreciate that the habitat in the Game transforms their disorientation into a technique that leads to knowledge. They embody a form of intelligence that would have been considered avant-garde in the twentieth century, but is now destined to become the most widespread form of mass intelligence. Like all elites, they can be either sublime or grotesque, or even both at the same time. However they behave, let it be clear that these elites are the ones who will end up laying down the laws of the Game—the invisible and therefore crucial laws, such as what is beautiful, what is right, what is alive, what is dead. If anyone had hoped that the digital insurrection would give us a world of equals where each and every person was responsible for creating their own value system, they need to think again. All revolutions create elites, and the elites are the ones who decide what happens next.

Already today, those capable of post-experience have emerged from the group and are standing there in front of us, clearly visible in a particular light. They have very recently—there is no turning back—become models, reference points, heroes for some. It hasn't been the smartest critics or opinion-makers asserting their status as models but, rather, the population of the Game. While writing this chapter, I happened to go to the main railway station in Rome, which everyone in the city passes by sooner or later—brilliant inhabitants of the Game, as well as those who are hanging on by the skin of their teeth and those who never really made it to the first square of the board—and where I saw an emblematic sequence of enormous billboards. The advertisements (gigantic portraits of young models) lined the access to the tracks with a solemnity that reminded me of the procession panels in the Parthenon frieze. The sequence of portraits—technically perfect, and the models are all as attractive as one would expect—is one of the most revealing expressions of post-experience I've ever seen. The billboards advertise a famous fashion label, the clothes the models are wearing, which they are presumably trying to sell. However, it's hard to see the clothes, because all I can see is what the company is rather brilliantly trying to sell in addition to the clothes: that is, a precise definition of a certain way of being in the world, at least for the elites of the Game. For every shot, every portrait, there is, potentially, a brief caption. I scribbled them all down in my notebook

✓ Has passports for two different countries and lives in neither.
✓ Has acted in first short film but doesn't want to boast about it.
✓ Loves doing yoga at dawn and sleeping later.
✓ Knows about stocks and shares and would love to know more about art.
✓ Convinced vegetarian, almost always.
✓ Loves New York, misses home.
✓ Has founded a successful advertising agency but always has time for a friend.
✓ Doesn't like being defined as an influencer but likes influencing people.

- ✓ Paints nudes as if they were landscapes, doesn't own a smartphone.
- ✓ Interior designer in Sao Paolo, goes climbing north of Rio.
- ✓ Owns electric cars and toothbrushes, washes dishes by hand.
- ✓ Gives wrong directions to tourists, later regrets it.
- ✓ Used to go out every weekend, now spends them in a country house.
- ✓ Resolves to go to bed earlier, starting next year.
- ✓ Inherited father's business, rejected a wardrobe that was a family heirloom.
- ✓ Left job at a bank to work as a baker, has never regretted it.
- ✓ Doesn't believe in horoscopes, is a typical Sagittarius.
- ✓ Accountant by day, tango dancer by night.
- ✓ Mistaken for an actor, prefers being behind the scenes.
- ✓ Works in digital publishing, still reads books.

Needless to say, the characters are all young and attractive. Needless to say, they represent a checklist of ethnic identities. Needless to say, they are dressed like gods on Earth. Needless to say, they embody individualism. Needless to say, they do not appear to have bosses. Needless to say, one would like nothing better than to send them all to hell. The fact that they are there, exhibited in that fashion in a railway station where commuters—whether they are in new, high-speed trains or old, local trains—are all desperately trying to carve out a halfway decent life for themselves, cries out for vengeance. It makes one wonder where the advertiser's sense of shame has gotten to. At the same time, needless to say—and it's important to understand—the entire gallery of portraits hits the bullseye with a level of precision that only the fashion business could aspire to. It interprets and pins down the type of people we all feel are emerging as the new elite: people who have learned a series of movements and skills from the digital habitat, which they have channeled into behavior in their analogical life in a way that has little or nothing to do with the digital world. Of course, these figures are caricatures because they are advertisements, but they are caricatures of winners. They are the ones who race ahead in the Game and who are impossible to catch up with, who reinvent logical figures that until the day before were oxymorons, who have built up their own constellations of meaning by putting together pieces of worlds that were previously distant, who use technology without being enslaved by it, who pass through the world lightly and peacefully, who bring the past into the present (bakeries!), who tame the present (how come they all have jobs?), who embrace the future (electric cars, what else?). They aren't nerds; they aren't engineers; they aren't programmers; they aren't dot-com billionaires. They are the intellectual elite of a new, vaguely humanist species that has replaced hard work and discipline with a capacity to connect the dots. The privilege of knowing has melted down and become a privilege of doing; the exertion of in-depth thinking has been turned upside down and become the pleasure of fast thinking.

Take this gallery of heroes, strip them of their commercial purpose, dust off their useless glamor, add a gloss of respect for others, and apply this characterization to people who are dealing with the real meaning of things rather than trying to sell cardigans. You will find the new elite: experts in post-experience who dismissed the twentieth century only after they had cleared out all its stores. They are the top digital natives of the Game; the ones who are translating every aspect of current knowledge into a different kind of knowledge, based on the surface of things, on mass individualism, on movement, and on lightness. OK, let's not show too much enthusiasm for this new elite. Many of them run around on wild goose chases, I know. They ride the surface of the Game at breakneck speed without gaining any traction whatsoever. In their depressingly narcissistic endeavors, post-experience is a perfect cover for people who are incapable of producing ideas or who are inadequately equipped for bearing the weight of intellectual honesty. These specimens remind me of those erudite thinkers who had a certain degree of success in the days of the twentieth-century elites. In their case, knowledge dressed up their ideas or veiled the paucity of their thoughts. For the new elites, on the other hand, it is speed together with an appearance of brilliance—a form of intensity. However, I am still convinced that, just as the twentieth-century elites comprised extraordinary and spectacular models of intelligence, the elites who have emerged from the Game crystallize around singular cases that have become more and more frequent—of prophetic, solid, and highly practical intelligence. People who did not contribute to inventing the Game, but who nonetheless know how to play it and therefore endow it with meaning. They are to the digital insurrection what Federer is to tennis. Not only do they keep the balls in the court; their serves are off the scale. These serves are language, in the noblest sense of the term. They are the runic alphabets that will help people recognize our civilization in ten thousand years' time.

MINOR POLITICAL FORAYS

The interesting anomaly of the Five Star Movement (M5S)

There's an interesting and surprising fossil that needs examining: traces of the insurgents' assault on the seat of political power. Smatterings really. Until now, only one limited case has been recorded and that was in Italy, a small, fairly irrelevant country when you think about it. In some ways, it was a country that was singularly unsuited to an experiment of the kind; it would have been more likely in one of the northern European countries where a centuries-old tradition of direct democracy, combined with entrepreneurship in the field of digital innovation, would have made it more natural. But no. The Five Star Movement (M5S) came about and won elections in Italy, a country that is not very digital, and that has a pretty baroque idea of political power and a penchant for the liberal arts over technical and scientific subjects. It's hard to grasp, I know.

Anyway, it happened, and there is one thing that we can learn from the success of M5S. This is that the twentieth-century idea of a political party—closed, hollow, static, everlasting, with mass membership and a bloated bureaucracy—is unsuited to the rules of the Game. It is *clearly* a leftover of a previous civilization. A party of this kind may continue to make sense as long as politics continues to operate within a fortress where the Game has no role. However, as soon as politics becomes a game that is open to other players (not necessarily digital natives, the other players may well be xenophobic populists or movements that bring people together under specific causes or banners), the twentieth-century idea of a political party begins to look like the Maginot Line, destined to be breached by enemy forces. A similar lesson can be learned from the experience in Spain of Podemos or tailor-made parties formed in a few months, such as Macron's in France. The visionary element of M5S was that they understood the inertia, but they were convinced they could use it in their favor. And they did, with undeniable determination and daring. I'm not really in a position to contribute an opinion on digital democracy (the possibility of voting with a click), nor am I that passionate about the subject. And yet, behind this experiment, there is the following important intuition: if you can't find an alternative to the twentieth-century party today; if you are unable to manipulate masses of mobile, changeable, unstable voters; if you want to catalyze the flowing currents without creating a bottleneck, such as a requirement for enrollment; if you don't know how to do all these things—you have no chance of winning an election.

> In a certain sense, this lesson should be extended to other institutions, which have so far been left undisturbed by the digital insurrection and have, therefore, continued to hibernate. In particular, education. It is possible that one of the problems with the education system today is its fixity; its permanent structures; the twentieth-century organization of time, space, and people. It could go on like this for decades, but the day someone decides to change schools, the first things to go will be classes, subjects, subject teachers, the school year, and exams. Our education system is a monolithic structure that goes against every inclination of the Game. Trust me, one day the whole edifice will be torn down.

There is another important lesson to learn from the Italian experience with the M5S movement. It is disappointing and derives from the political agenda M5S proposed to its electors. Against all logic, the agenda is in many ways twentieth century in spirit. It is hard to recognize the hobbyhorses of the digital insurrection in the program. For example, they are anti-Europe and do not exclude the idea of leaving the Euro Zone. They sympathize with Brexit. They support the concept of a job for life. What do these things have in common with the idea of an open playing field, the cult for movement, and the vaguely hippy idea of a shared world? Nothing at all. Likewise, their position regarding immigration: they want to keep the gates closed and maybe even put a lock on them. Their taste for sustainable degrowth is also slightly suspicious given that the DNA of the movement was supposed to be the same as that of the fiercely ambitious pioneers of the digital movement. It's weird: it is as if M5S were digital without really being digital. The most evident symptom of this anomaly is what is happening as I write these pages here. M5S is forging a political alliance with a populist, xenophobic party that would have been called right wing in the past (the League), supported by small and medium-sized firms in the north of the country. Its members are hard-working, no-nonsense folk with no taste for poetry. They are traditional, pragmatic, rough-and-ready in their logic, with huge trust in the past, and unyielding to the sirens of the future. In sum, they are solid in a completely old-fashioned way. What do these people have in common with a political movement that was born from the digital insurrection? In theory, absolutely nothing. They should be irreconcilable at an anthropological and cultural level, never mind a political level. And yet, there they are in a government coalition. They understand each other, they even have shared objectives. That's what I fail to understand under the circumstances.

Well, obviously politics has a different set of rules, and there are many reasons—some of them the lowest imaginable—why this partnership may work. I accept that. However, there is still an anomaly, and in a book like this, a lesson must be drawn, setting aside the squabbling, the base political bargaining, and the run-of-the-mill power struggles. I will thus try to look at the experiment from much higher up, attempting to forget that Italy is my own country. Things become a little clearer from up here.

I can see at least two points where the digital insurrection and right-wing populism may overlap and recognize each other. One is the visceral hatred of the elite; the other, their instinctive inclination for mass egotism.

I don't want to tackle the issue of populist movements here. Let's stay focused on the Game and on what a phenomenon like the M5S has taught us. What it has taught us is that the Game generates solid social and mental structures that can unleash disruptive impulses. For example, there was that old idea of diminishing the role of the elite, of freeing the world from the unjustified power wielded by those who held the keys to knowledge, and of giving everyone both the right to direct contact with the world and the duty to make their own decisions. The idea was a direct consequence of the disasters that took place over the course of the twentieth century. It wasn't a bad idea in itself, I would say. However, it is unfortunately equally likely that, in a later simplification, everything will be boiled down to a hate-fueled showdown, an indiscriminate, nonviolent manhunt whose only aim was to punish the elite, who were seen as failures and accused of occupying positions of responsibility for no justifiable reason. In general, the inhabitants of the Game do not seem to be fanatically attracted to simplifications of this kind. The Five Star Movement, for example, is full of people who are far more interested in the challenge of governing a country than in trying to kick all the current politicians and commanders out of the system. And yet, the experience of the M5S shows that there are situations where simplifications of this kind can be irresistibly visited on the masses, and politics is one of these. In politics, emotivity reigns supreme. Existing mental frameworks are swept away by the unadulterated flow of collective impulses. It may thus happen that a digital approach to the world is reduced, under certain circumstances, to an instinct, a sign of intolerance, a generalized "fuck off." This is the instant when right-wing populism is contiguous with digital democracy. It doesn't necessarily mean anything, and it's not even that important, but for those of us studying the Game, it is significant in itself. The lesson is that the Game has a gut, and every now and again, gut feelings take over. When they do, any kind of disbandment is possible and can be expressed through regression, old-fashioned anger, or dancing arm-in-arm with the right wing.

Similarly, if for years you have been cultivating mass individualism, you are a step away from triggering an undesired effect, which is mass egotism. That is, the fact that millions of individuals have become incapable of looking ahead and predicting the successive twenty moves of the Game. Rather, they cling blindly to their next move, which is the one that defends them—precisely them, and only them. I don't believe there was this level of egotism in the fathers of the digital insurrection. There was a great deal of individualism, of course. Perhaps too much. However, egotism was not in their genes. They were far-sighted, and they reasoned in terms of a community with an instinct not to leave anyone behind. Nevertheless, if you cultivate augmented humanity, fertilize individual egos, and nurture mass individualism, you certainly run the risk of slipping into a form of mass egotism at any moment, and for years to come. All it takes is a context where things are hard, where there is a light breeze of fear or a gust of emotivity. All you need are huddled masses of migrants at the gates, and it's done. That's when you find right-wing populists marching at your side. It doesn't necessarily mean anything, and it's not even that important, but for those of us studying the Game it is significant in itself. The lesson is that the Game has a gut, and every now and again, gut feelings take over. When they do, any kind of disbandment is possible, which can be expressed through regression, old-fashioned anger, or dancing arm-in-arm with the right wing (I know, I know, I wrote it already. I was just doing it to underline the symmetry of it).

To sum up: the Game stuck its nose into politics, in an insignificant corner of the world admittedly, but nonetheless, it did so. This experiment has taught us two lessons. First, the fate of twentieth-century parties is to lose against any political body that is more fluid. Second, the Game has a gut, an irrational well of sentiment; it is not merely technique, rationality, and efficiency.

Let us set these two fossils aside for now (carefully now, they are valuable!).

THE REDISCOVERY OF EVERYTHING

As we all know, when Brin and Page went to speak to their professor at Stanford in order to propose a research project that later turned into Google, the first thing the dear old professor said to them was, "Fine, but you'll have to download all the web pages." The professor must have felt that his objection had served as a red light for the project, given that there were 2.5 million web pages in the world at the time. The two fresh-faced students didn't blink an eye. "Where's the problem?" they quipped. That was the moment a way of thinking was launched that was later shared by all the organisms born of the digital insurrection. That is, considering EVERYTHING a reasonable challenge, a reasonable playing field, or rather, the only playing field worth playing on. Amazon trumpeted itself immediately as the biggest bookshop in the world because it was able, in effect, to retrieve every book in the world (or at least the ones written in English, as Americans find it hard to accept that other people exist). eBay potentially reached everyone in the world, as did email. What must have seemed immediately clear was that, once world data had melted into an extremely agile, immaterial form, the furthest corners of the world suddenly became visible to the naked eye. The idea that these distant regions could be explored was no longer a pioneer's epic vision; it was simply a normal gesture involving little more than patience and dedication. Anyone who wanted to download every single web page in the world could rent a garage, fill it with computers, and do so. And that was it. Similarly, once music was digitalized and you could lie on your bed, conjure it up instantly, and listen to it through your device, why would anyone limit themselves to just classical, Occitan, or 1960s music? Let's just digitalize all the music in the world, and then I can choose what I want, right? That's much better.

In the past, EVERYTHING was the name we gave to a hypothetical magnitude; since the beginning of the digital insurrection, however, it has not only become the name of a measurable quantity that can be individually possessed but also, in the long run, the name of the only quantity present on the market. It is, in sum, the only significant unit of measurement. Take Spotify, for example: all the music in the world. What is revealing in that particular vertebra is not so much the fact that it actually contains (nearly) all the music in the world; it is THE WAY THE SERVICE IS PAID FOR. We don't pay by the piece, we pay for access to all the music in the world. In the most evident way possible in this case, there is only one thing that you can attach a price tag to: EVERYTHING. Everything becomes a commodity. The only one. A transition of this significance must not be underestimated. It is pure revolution, with enormous consequences.

The first was cultural, or mental perhaps. If EVERYTHING becomes a unit for measuring things—the epic aim for every company and every sales pitch—you instantly create a victim, which is INFINITY. If you can get to the bottom of EVERYTHING, there can be no infinity. It's a good idea, however, to remember that, as it happens, infinity was one of the pillars of romantic sensibility, the soil that nurtured twentieth-century culture. Which brings us back to the blood feud we started with. Digital insurgents had an enviable aim, with the infinity pillar their target: they took aim, fired, and the pillar crumbled. It was a good idea in principle, because there is no denying that a certain poetic culture of infinity in the twentieth century had bred a form of irrationalism, not to mention mysticism, which was not so strange given the madness of the time. Exploring the territory, reclaiming the land, and converting its inhabitants to less risky cultures was a valid mission. Thousands of apps are doing precisely this: they are annihilating the concept of infinity and reducing the uncontrolled margins of the world to a bare minimum. This is a tiny example, but when you have an app on your phone that gives you the lyrics to every song that has ever existed, what ceases to exist are the confines between the songs you know the words to and the infinite number of lyrics you don't know. What disappears is a pause, a time-lag, a vacuum, a shadow—the perception of an infinity you are not able to inhabit. Considering that, clicking on the icon right next to this one, you can knock down all language barriers by translating anything you like from any language in the world simply by taking a picture of the source text, the physical perception of a world where anyone can reach the outermost edges of their experience starts to become more insistent. If you like, you can just go on clicking: hopping from Google to Wikipedia and then on to YouPorn, you will only find complete worlds where extremes have become the rule and where EVERYTHING is a reasonable measure of something you are perfectly used to. Multiply this sensation hundreds of times a day for days and for years, and you will see that in the Game, infinity is a category in disuse. It survives as a concept that is considered a little kitsch, something that might provide entertainment for a bargain-seeking audience. In the rest of the world, there is a dominant form of technical rationality with an immense capacity for calculation, and thus an inclination toward imagining that there are no longer any real outermost edges of the world that cannot be reached. Again, the models appear to be video games. Only a very few players reach the highest levels, but everyone knows they can be reached eventually. There is no such thing as an unachievable infinity. Similarly, TV series may feel like they are infinite, but they are not. There is simply never a last episode. If, in the pilot, the screenwriters told you they had no idea how things were going to end, you wouldn't be happy about it. You may get fed up as the series goes on, but when you start watching, you need to feel that there's an end somewhere and someone knows what it is. This is the way, tool after tool, that this peculiar strategy has been implemented, becoming one of the pillars of the Game, one of the things that hold it together. What happened is that everything in the world was stored in colossal warehouses that eliminated the unknown quantity of infinity. At that point, everyone went to live there, protected by high walls that would never be reached, but that they knew were real.

This, of course, takes all the mystery out of creation. It may be what caused, for example, the fixity, non-resonance, and lack of vibration in the products of digital culture we described before. Without the resonance of some kind of infinity, any form of reality would sound a little wooden. In the same pages, however, we commented that, owing to the post-experience technique, the Game has succeeded in integrating the system with a beloved vibration, a touch of mystery, perhaps a meaningful version of infinity. What seems to be locked into a self-sufficient EVERYTHING can be liberated if you link all the various warehouses and use them as a strap system to propel you on a journey that can be truly infinite: that is, post-experience. The journey changes something; it opens up the world again. It is no longer all-embracing.

So, what we have in sight now is an articulated strategic model that we need to pin down because, as we have said, it is one of the pillars of the Game. The strategic model has five steps:

4. Archive everything in the world in immense warehouses that eliminate the unknown quantity of infinity;

5. Move there, protected by walls that can never be reached, but that exist;

6. Recover infinity by linking all the warehouses;

7. Give the keys to everyone;

8. Live anywhere.

If you put these five steps in a sequence, they form the classic opening moves of the Game.

Understanding this game strategy helps us comprehend the second sequence that the rediscovery of EVERYTHING has imprinted on our way of being in the world. This is important because it relates to the business world and, even more significantly, to a certain idea of competition and pluralism. Let's take a look.

As we have been able to document examining that valuable archeological artifact that is Google, a strong urge to consider EVERYTHING as the only real unit of measure meant that the protagonists of the digital insurrection tended themselves to BE, in their turn, EVERYTHING. What I mean is that Google is not just a search engine, it is THE SEARCH ENGINE. It has no significant competitors (at least in the West) and nobody really expects it to have any. This instinctive and inexorable occupation of space—of all available space—has established a business model that can easily be spotted in many of the vertebrae along the backbone of the digital insurrection: A GOOD BUSINESS IS A BUSINESS WITH ONLY ONE PLAYER, YOURSELF. I doubt whether Henry Ford would ever have dreamed of a business founded on this premise (and he was quite a mythomaniac), not to mention Walt Disney (another businessman who was paranoid about controlling the market). In the digital era, on the other hand, the model seems perfectly reasonable. In fact, nobody really questions why Amazon, Facebook, or Twitter have no real competitors, while companies like Nestlé and Volkswagen do. Something has changed, and in order to try to explain what, I'll use a card game metaphor. In the past, doing business meant inventing games that could be played with a pack of cards that already existed; the winner was the one who invented the best game. Nowadays, doing business means inventing a pack of cards that did not exist before and that you can only use for one game. The one you invented. Period.

It doesn't always work, or we wouldn't have Apple and Samsung fighting to sell us their latest smartphones or Safari and Google Chrome competing for dominion of the web. There are lots of different makes of tablets, and the duel between Microsoft and Apple is ongoing. And yet, WhatsApp, for example, or Twitter, Google, Spotify, and Facebook have little or no competition. In some cases, their names even become verbs to express the action of using them. This reveals something important about the civilization that created a world of this kind: the twentieth-century concept of pluralism (where many different players use the same game board) was alien to them. If anything, they thought pluralism complicated things for no reason, wasted energy, and potentially created chaos. Rather than sapping their energy managing the many different players on the same game board, they preferred to use their energy to multiply the number of game boards they could play on. Their idea of efficiency was one player per game with an innumerable number of games. They were convinced this framework would protect them from creating monopolies or concentrations of power, that it would avoid the horror of orthodoxy or the fear of Orwellian groupthink. I know, it's weird to say this when you look at giants such as Google and Facebook today. And yet, at least in the early days of the digital insurrection, they believed that the only way to truly free citizens was to provide them with a range of game boards rather than one game board crowded with a range of players. They weren't the kind of people who worried about things like ensuring that every political party had an equal amount of TV exposure; they simply created the conditions for every party to have its own TV channel. Digital television, with its plethora of channels, is precisely that. It's hard to deny that it works.

Allow me to indulge in a little story from my past. I grew up in the sixties with only one state-owned TV news program. We would listen to the evening news during dinner, not in religious silence but with a certain amount of respect. There were no other channels. Only one newspaper ever came into the house, and it was always the same one, owned by the richest man in my town (and in the whole of Italy, I think). I was young enough not to take into account the fact that adults can lie to you. At the table one evening, after I'd finished my pasta and was tucking into my veal cutlet, I heard the news anchor—who, for me, was the equivalent to a god on Earth—discussing a war I knew next to nothing about in a faraway place called Vietnam. The question is, did I have any chance of hearing the truth—even a half-truth—about that war? Absolutely not. In my world view, the tall, healthy looking, white teethed Americans were the good guys, while the little Vietcong with rotten teeth were the bad guys. That was the sum of it. Was there anything in the information supply system I grew up with that might have freed me from my blindness and the dark ages? No, there wasn't. My country decided to remedy the situation by launching two new TV channels. They were emanations of the two main political parties, and they were added to the single channel that was in the hands of the governing party. The result was that there were three TV anchors—three divinities—who split the world picture into three different world views. The Vietnam War was almost over, but if it had continued, the first channel would have announced an American victory, the second would have debated the chaos, and for the third, the Vietcong had already won the war years before. The remedy was worse than the sickness, as I'm sure you can appreciate. There was only one solution: to create a system whereby the news reached you from all sides, on many different devices, in diverse habitats, none of which are sacred, all of which are to be taken with a grain of salt, and most of which are produced by as many authorities as possible. The main aim was to ensure that it was no longer the elites providing the news, paid for by the rich and powerful of the planet.

Well, this is exactly what we did.

My son is now the same age I was when Ho Chi Minh was giving the US a hard time without my knowing anything about it. Whatever way you frame the argument, there is no way to convince me that he would be better off (or better informed) in a system with three TV news broadcasts and one newspaper (belonging to the richest man in the city) compared to what he has at his fingertips every day in the Game. I understand the risks. I share the concerns. I respect the idea that we must be critically vigilant. But I still believe that the system today gives him a greater chance than I had fifty years ago to become an informed, mindful, and mature citizen. In light of this conviction, I would advise caution in tackling the issue of the giant monopolies today. I would even go as far as suggesting that the problem has been overstated in a classic twentieth-century reflex that does not take into account the current state of the playing field. It's a little like stepping out of your house terrified you'll be run over by a horse and carriage. I suspect a fear of this kind is a little obsolete. It has been made obsolete by the fact that the monopoly you dread gives you access to a world where movement is idolized, the multiplication of habitats has become a religion, transversal moves are the official gait of its inhabitants, and no building is inhabited if it is not linked to a strap system of other buildings. I'll sum it up, making it as annoyingly simple as I can: in a world with Google, the Google monopoly is not so dangerous. In a world with Facebook, the fact that Facebook is everywhere is not so worrying. In a world where four hundred hours of video are uploaded every minute onto YouTube, the fact that YouTube exists and is basically a monopoly is peculiar, but it's not a tragedy.

Google. Facebook. YouTube. Try to imagine them in the twentieth century under Nazism or in the Soviet Union. Those were real tragedies.

I have news for you, though: the twentieth century is over.

The question to ask is this: has the ecosystem of the Game —which has a high tolerance, if not need, for monopolies— developed antibodies in order to stave off the degeneration of a playing field blocked by four or five players?

Good question, eh?

I'll attempt to provide an answer to this question in the final chapter of this book, which I will call "Contemporary Humanities," using an expression a couple of my students suggested to me.

THE SECOND WAR OF RESISTANCE

There is a very clear clue we cannot ignore here: it was at the height of the Era of the Game when a *second war of resistance* was sparked. If you remember, the first fairly unsuccessful wave of resistance was triggered in the nineties and ended up going underground. Starting in 2015, I would say, a flame was kindled and then fanned by the Brexit referendum in the UK and Trump's election victory in the US. Both of these events signaled that there were potential deviations of the Game, which opened people's eyes to the dangers. One interesting feature of the second war of resistance is that the partisans were not only veterans of the first war (who were still stubbornly twentieth century in their approach), there were also many of the Game's offspring, some of them even émigrés from the new elite, individuals who had fought *for* the digital insurrection, not against it. What makes them rebels is the feeling that the system is degenerating. They are not fighting against the Game; they are fighting *in the name of* the Game, in the name of the values it once represented.

The countermovement is so fascinating I decided to try to find out more about it. The result is that I now more or less know what these rebels can't accept. What makes them angry. I'll try to summarize it in a few clear points.

1. The Game was born as an open playing field that redistributed power but was slowly preyed upon by an infinitesimal number of players. They grabbed practically everything and often made alliances among themselves to stitch things up even further. We are talking about Google, Facebook, Amazon, Microsoft, and Apple. The same old names.

2. The richer they get, the more these players can buy everything up in a vicious circle that will continue to increase, bolstering their immeasurable power. The most dangerous part is that they are also getting their hands on all the innovation; that is, the future. They grab all the patents and are the only ones with sufficient financial resources to invest in artificial intelligence (AI).

3. A large portion of their profits is gleaned from a random (or perhaps cannily conscious) use of the data we leave on the web. There is a systematic violation of our privacy, which seems to be the price we are forced to pay for the services these players make available to us free of charge. This appears to be the rule of thumb: when the service is free, what they are actually selling is you, the user.

4. Another portion of their profits is garnered from a deceptively simple concept: these guys don't pay taxes. Rather, if they

do pay any taxes at all, they are nowhere near as high as they should be.

5. The circulation of ideas, news items, and different truths has become a market in itself, dominated again by very few players. The suspicion that they could easily use their dominant position to manipulate our beliefs is well-founded. They probably already do.

6. Whatever the original intention, what the Game has produced is a vast chasm between those who are suited to it and those who are not, between rich and poor, strong and weak. Traditional capitalism—even at the height of its golden age—had never distributed dividends in such an unequal, unfair, and unsustainable manner.

7. By dint of handing out content at a laughably low cost or free, the Game has committed a genocide of inventors, innovators, and professionals. The work of a journalist, musician, or writer has been transformed into merchandize that floats on the web, creating profits that vanish into thin air and almost never end up in the hands of the original author. Distributing, rather than creating, wins you profits. If this goes on, year after year, you'll have to go to the other side of the world to find someone willing to create anything.

8. The more games designed to solve problems are perfected, the more one begins to wonder whether perhaps these games have a vaguely narcotic effect on the general population. The Game can use this effect to keep the weaker members of society docile, drugging them just enough to make them indifferent to the fact that they are basically servants.

As you can see, this is no joke. These objections are scary.

However, I think we need to stay focused and go back to being archeologists, noting three things.

First, none of these objections would ever have emerged in the nineties, because they are the consequences of the Game. They are the symptoms of a malaise generated by the more recent developments of the digital insurrection. They are not a regurgitation of twentieth-century culture; they were brought about by the culture of the Game. Second, these objections do not challenge the premises of the Game. Rather, they hypothesize that there has been a perverse and unforeseen distortion in the way it was used. As often happens at the tail end of a revolution, the accusation was of betraying its original ideals. The third thing is crucial but irksome to note. That is, there is a quite large, irrational component to many of these objections. Many of the fears are based on hearsay: such and such *may* or *might* happen. Trust me, the objections are all perfectly valid, but if you approach them with no bias and with a true desire to look at the facts with a clear eye, you will soon realize that things are not what they seem. Your desire to vent your anger is probably greater than your arguments for doing so.

The fact is, since a certain point, there has been a growing need to distance oneself from the Game, to put one's foot down. This sensation does not necessarily depend on facts; it is more like part of the unstoppable movement of a civilization that is trying to recover its balance after finding itself leaning out too far into the future. It is as if these human beings needed to find a fault line in the system in order to slow it down so that they could catch up. I'll venture one more thing. It is as if they have a spasmodic need to find a scapegoat, a bad guy, maybe in order to put their minds at rest that not everything is bad. The resentment they feel against the big players seems to have reduced to zero the chance that they'll remember they were the ones who chose this world and who helped create it. They are people who use Google a lot and hate Google; people who can't live without WhatsApp and see Zuckerberg as a devil in the flesh; people who have an iPhone and believe iPhones make people stupid. The vast online news agency they get their information from chastises the big players on an almost daily basis, and then a window pops up coincidentally after the third news item, advertising an extremely rare make of vacuum cleaner they just happen to have been looking into a few weeks before on a search engine. Well-informed people are shocked by the fact that if you have neo-Nazi leanings, you will get suggestions in a column to the right of your screen on YouTube which will very probably add fodder to your particular penchants. Well, what do they think YouTube would suggest? Speeches by Martin Luther King? If they made the same suggestions to us, providing us with videos of delirious rants on white supremacy, would we consider it a good sign of YouTube's objectivity? Is the fact that the web gives you only the news you want to hear—thus reinforcing your convictions—something that people who have experienced church parishes, party branch offices, rotary clubs, the TV news when there was only one channel, or 1960s newspapers seriously fear? I'm saying all this—I beg you to understand—not to deny that the objections I listed above are legitimate or even justified. Rather, to explain that adhering to these objections is often a sign of blind, instinctive, irrational, disproportionate fear, which is also, of course, a tremendously real, physical, animal sentiment. It is a significant symptom because it reveals the fact that, in the advanced Era of the Game, an almost pathological dependence on the *tools* of the Game has grown hand in hand with an urgent, almost physical rejection of the *philosophy* of the Game. It is a kind of controlled schizophrenia. The Game is out there, it is working, but the players are starting to hate it. They are technically in agreement but mentally in dissent.

While all this has been happening, another energy force has been at work which needs mentioning, even though it makes things even more complicated. It is not a resistance movement. It is another phenomenon that feels more like a mutiny, resulting from the fact that many of those who have been rejected, unrecognized, disappointed, or exploited by the Game decided to mobilize en masse. They are a completely different group from the rebel elite who lament the betrayal of the original ideals of the insurrection. They are the ones left behind by the Game. The novelty is that they have simply stopped playing and are digging in their heels. They have done this in a fairly bizarre way, which I can only describe to you with the example of Trump and what he represents. If you think about it, there is a kind of schizophrenia in the way he conducts his presidency. On the one hand, he is tweeting with world leaders rather than using twentieth-century diplomatic channels. There is even the suspicion that hackers—the guerillas of the Game—helped him get into office, though he didn't necessarily ask for their help. On the other hand, he is engaging in old-fashioned trade wars, and he dreams of building a security wall along the Mexican border. It's an odd way of going about things. It's hard to understand what he thinks he's doing, but it is easy to see that many people have been behaving the same way in recent years. These people elected the man as president of the United States. His way of being in the Game embodies that of a great number of people. One could call them mutineers: they use the ship, but they have occupied the engine room and taken a U-turn back to a safe harbor; they use the Game, but they are trying to convert its ideals, thus betraying the reason it came about in the first place. They have split the mental revolution off from the technical revolution. In short, they have walked into the game room, grabbed what they wanted, and set fire to the rest.

It's a little worrying.

What we can surmise from studying the archeological ruins

of the Era of the Game at its zenith are signs of an incredibly harsh conflict. Squeezed in a pincer movement between resistance fighters and mutineers, the Game looks like a regime on the brink of collapse.

Is it really?

What fascinates me is that the answer is no. The Game has had some quakes. It is the object of riots and unrest, it spews out paradoxes we have no idea how to deal with, but you have to ask yourself: is there really a sensible, reasonable, intelligent alliance out there that wishes to dispense with the Game altogether?

No, there isn't.

There are more and more tools available, people are increasingly skilled at using them, there is greater vigilance regarding their adverse effects, and there are more highly refined techniques for offsetting their collateral effects. A civilization that wanted to turn the game table around wouldn't be developing all these checks and balances. It looks to me very much like a civilization that wants to go ahead, not bail out.

So, what is really going on in the gut of the Game? What travails is it traversing? Why is it writhing in agony? What is the point of splitting the consciousness of its inhabitants in two?

What shall we call all this on our third map?

MAPPA MUNDI 3

To sum up, the insurgents went ahead and followed their chosen path, settling in the promised land after the mass migrations of the twentieth century, where the Game had become more than just a technique, idea, or magic trick for the brightest and best: it had become a homeland for everyone.

Several years went by with minor quakes here and there which were, nonetheless, not without significant consequences. The *human-keyboard-screen* posture was gradually rarified to a POSTURE ZERO, whereby devices became a quasi-organic extension of the human body. When apps started to proliferate and the amusing idea (that could have featured in fairy tales) of shifting data onto clouds was first aired, the hard confines between this world and the otherworld definitively melted away. Technology allowed people to go back and forth so fast that reality became a single system with dual pumps, just as the digital insurrection had envisaged in its dawn. The idea of a *real* life as distinct from an *artificial* life contained in our devices similarly melted away in the common perception that life was a single, giant game board that was open and accessible to all.

The best way to exploit this situation was to develop the specific ability to plough the seas at high speed, capturing in one's sails the meaning of things as it surfaced, tacking and jibing, drawing figures in the water which one knew would become concepts, ideas, works of art, products. It was a sport that had never been developed before, which we described as post-experience. People soon discovered it was tough. For this reason, silently but inexorably, entirely new elites began to form. They were very different from their twentieth-century counterparts, with very different skills. There was one thing they were very good at, however: that particular sport. In the reign of post-experience, they were like gods. The Game might have been envisaged as a world without elites, but that was not how things developed. Quite soon, a group of highly skilled adepts formed and began to establish models, amass great wealth, impose their tastes, and set rules. The archeological remains we've been able to examine do not allow us to establish precisely how far this new caste will go in establishing their dominion. There are evident signs, however, that the caste is there, solidifying its hold on the Game, as we can see from the ridges on the ground, which reveal an unintended, perhaps undesired, effect that was certainly not pursued.

It is not the only effect, of course. The site we have excavated is full of fossils which tell the story of the many inconvenient collateral effects the Game had not taken into account. The most evident is that the simple desire to give every human being the opportunity to have a computer on their desk at home—a pull factor for the masses on the social periphery who were attracted toward the center of the Game—had the exhilarating result of sweeping away existing barriers of class and culture, and giving rights and dignity to vast swaths of the population. At the same time, however, these same citizens had the dubious pleasure of discovering that the backbone of the Game was not necessarily strong enough to bear the weight of all that extra muscle. One result was that what we called augmented humanity, boosted by the proliferation of affordable devices, sowed the seeds of a renewed self-awareness which, in its turn, transformed the social fabric into a paradoxical phenomenon we described as mass individualism. The oxymoron in the name makes it clear that the situation is not easy to govern; it represents, in any case, a shock wave that the Game had not reckoned with and had not yet acquired the tools to tackle.

Similarly, an exaggerated power of calculation, designed to feed increasingly demanding devices, has brought about a general impression that EVERYTHING is a known quantity and ultimately the only product that is worth buying and that is lucrative to sell. As we have seen, colossal monopolies are the result. Other collateral damage includes games with only one (solitary) player or businesses where there is only one chair. The fact that these phenomena did not develop on a planet with feet made of lead like ours was in the twentieth century meant that people rushed to interpret them as symbols of mortal danger. It is unclear, however, whether living together on the slippery dance floor of the Game is not equally dangerous.

At the end of the day, if we trust our observations of the

archaeological remains, we must face the surprising fact that, just as it reached its apex, the Game began to cave in, revealing cracks, instabilities, and underground collapses. From a certain point onward, we can clearly see it was besieged by the contemporary attack of three forces which, in theory, had very little to do with one another: veterans of the twentieth century who were not yet resigned; purists of the Game who wished to reclaim the original spirit of freedom; and those who had been locked out of the Game altogether—the unwilling, the excluded, the habitual losers. The funny thing, which will be duly noted on the map, is that all three of these forces, including the veterans of the twentieth century, launched their attack on the Game from the inside, armed with—nay, dependent on—digital tools. They did not for a second entertain the idea of going back to a pre-digital civilization. In at least two of these forces (the veterans and the losers), it would appear that what they were aiming for was to grab all the tools and abandon the Game. They wanted to exploit the technological revolution but defuse the mental and social consequences. Squaring the circle, I suppose you could call it.

Slyly, the Game let them play their own game, aware perhaps of the cracks but confident that these fissures were small details that would be brushed aside during the inexorable march forward of its model of behavior. The Game barely remembers that it came about in order to obliterate its destructive past. For a while now, it has presented itself as a civilization with its own justification and its own objectives. For many of its inhabitants, it is not the enemy; it is the world they are proud to have built. The more disruptive the opposition is, the more millions of human beings seem blindly determined to leave their homes every day to go out and build their little segment of the Game, convinced it is their homeland. They are already hatching the next plan without attempting to hide it: artificial intelligence will soon make the second war of resistance slide back into obsolescence. There will be more important things to discuss and more radical scenarios to deal with. We have already seen that nothing that happens is by chance; the seeds were sown years before in the open fields of the Game. Whatever the consequences of artificial intelligence, human beings started working on the technology years before: when they accepted making a pact with machines, when they chose to adopt a posture zero, when they digitalized the world so that it could be processed by immensely powerful calculators, when they privileged tools over theory, when they left engineers at the helm of their liberation movement, when they set sail for the seas of the otherworld, when they repudiated the elites who taught them to die, when they accepted the risk of an open playing field, when they chose peace and forgot about infinity. They sowed the seeds and harvested the crops. There will be many more harvests and plentiful produce—fruits people have never seen before—to mitigate the insidious effects of nostalgia, together with the eternal return of fear.

Here we are. Many pages back, I started collecting the footprints of these human beings with the idea of reconstructing their path and measuring how far they stood from both happiness and fear. I was thinking of drawing maps, and here I am leafing through them, looking at them, touching them, rereading the names, glancing at those beautiful lines that represent solid borders. There are many white spaces which have not yet been filled in; I've changed some of the altitudes of the mountain ranges and added a few details here and there. Like every cartographer, I know I have represented things as precisely as I could, accepting that it is, by definition, an inexact representation of the world. When you are mapping continents, you cannot render the color of the flowers or what people are feeling as they gaze at a sunset. Every map has its own interpretation of reality, one of many. The map I have been working on represents one thing only in the recent progress of mankind: the digital turning point. In order to understand these human beings, we could have equally examined the history of drugs, sport, or food. Although I have spent a ridiculous number of hours trying to understand the importance of the web in our lives, I am well aware that it would be just as useful to examine the role of Prozac, slow food, Pope John Paul II's theology, *The Simpsons*, *Pulp Fiction*, the Erasmus college exchange program, the arrival of sneakers, the disappearance of pantries, the craze for sushi, Amnesty International, MTV, Dubai, Bitcoin, climate change, or Madonna's career. Even the 1992 elimination of back passes to the goalkeeper in soccer tells us something about our society. We should all be willing and able to study everything, to draw all the possible maps, and to overlap them to see the results. It would be a typical post-experience stunt for the elites of the Game. Kids in junior high today who spend their afternoons playing *Far Cry* may be able to do it one day. I place great hope in them.

In any case, we've done a good job so far. If we went back and reread the first two chapters, they would feel almost like pre-history by now (don't actually do it, it would be too boring; trust me on this). We have made a lot of progress, though. There may be a lot of errors, but the path is now clearly visible, the logic is plain. A genealogy has surfaced, and the outline of a civilization can be seen poking out from the shadows. It's a great result, believe me. I may be overestimating my understanding here, but my feeling is that if my son asked me where we are going, I would now know what to say. Where we come from is already clear. Why we are doing this exercise is clear. Two hundred or so pages ago, if you remember, it was me asking him where we were going.

OK, good. We can tick that box.

I could stop here. You can stop here. And yet, as you can easily verify, there's a whole chunk of the book left. You really don't have to read it, but I had to write it. It's a matter between me and myself, a challenge if you like. The fact is that if you have drawn maps, you want to use them to navigate your way around the world. I harbored a desire to use them to help me find my way through two areas that fascinate me in particular: the area of truth and that of works of art. People are full of baloney when they talk about these areas, which really annoys me. In short, I felt like creating a bit of order using the maps I have drawn. It may seem a little arrogant on my part. You're absolutely right. It is.

There's a final chapter, too. This is called "Contemporary Humanities," which I think I already told you is not my expression. It came out of hours spent at the Holden School of Creative Writing trying to understand what we are teaching, what we are trying to teach, and what we *actually* succeed in teaching. We couldn't really get a handle on it until somebody younger than me, of course, came up with the expression "Contemporary Humanities." When I first heard it, I realized it did not only express what we taught at the Holden School, it also connected to the Game. In fact, it gave an unusually precise name to an area of the Game that was strategic and central but currently almost empty. It was at that precise moment that I discovered what the area where I lived was called.

This is why I've called the last chapter of the book "Contemporary Humanities." In this chapter, I will give my opinion on the Game, the digital insurrection, Steve Jobs, Mark Zuckerberg, and even the background colors of WhatsApp. As you may have noticed, I have so far tried to avoid standing in judgment. It's not out of shyness or cowardice, I can assure you. The fact is, when I am analyzing something, I waste too much time if I attempt to establish whether I like or dislike it or evaluate it in any way. If I am studying harmony in Debussy, it doesn't really matter whether I like his music or not. If I try to understand my kids, for example, I will make fewer blunders if I set aside my blind adoration of them. It's a methodology that I trust and that helps me achieve my aims. Therefore, as I discussed the web or Facebook in this book, I tried to keep my pleasure at bay or my dismay in check. In short, my objective was to understand, not to judge. It was not the time for a verdict.

At the end of the book, however, why not? I'll be happy to tell you what I think. You can take my opinions as screen credits rolling at the end of the film, if you get that far. That's kind of what they are.

Ah, I forgot. The Five Star Movement (M5S) ended up going into government first with the League—the populist, xenophobic party I was telling you about—and later with the center-left Democratic Party. I'm telling you about it because I promised I would.

So be it.

Username
Password

Play

▸ Maps: Seven Thematic Routes

- The Game
- Mass Individualism
- Posture Zero
- The Demise of the Elite
- Dematerialization
- Post-Experience
- The Rediscovery of Everything

Level Up

The Game

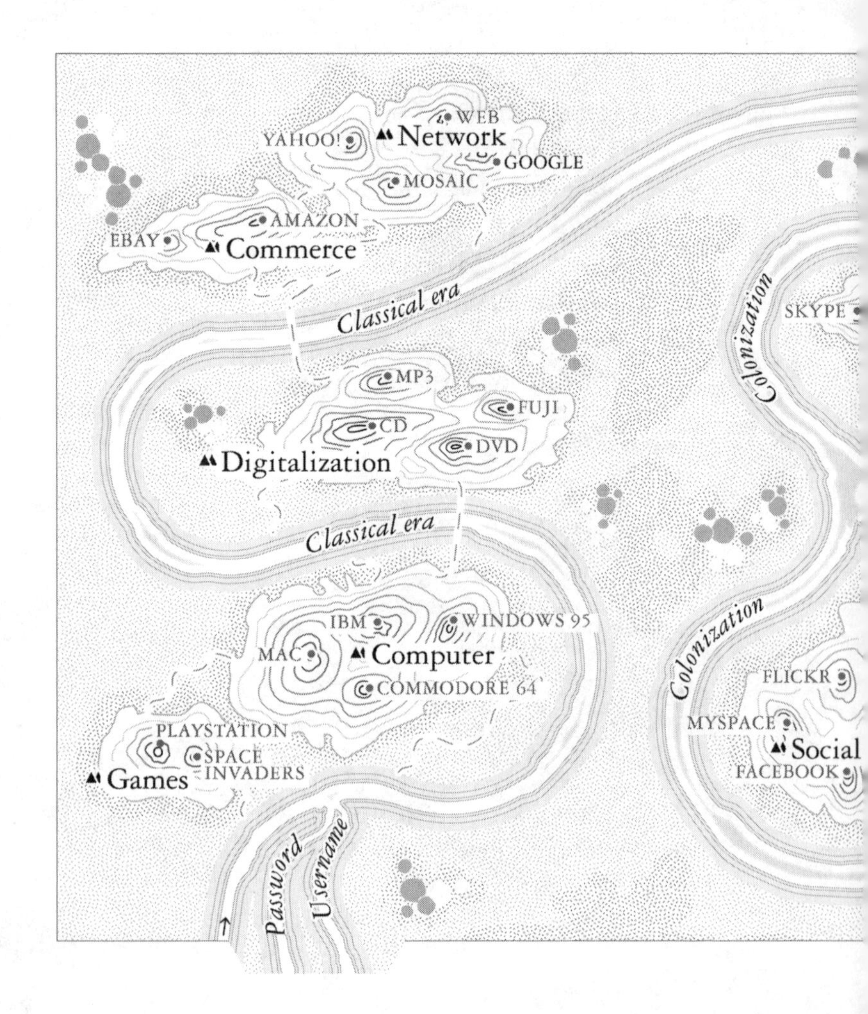

WEB
YAHOO!
Network
GOOGLE
MOSAIC
AMAZON
EBAY
Commerce
Classical era
SKYPE
Colonization
MP3
FUJI
CD
DVD
Digitalization
Classical era
IBM
WINDOWS 95
MAC
Computer
COMMODORE 64
Colonization
FLICKR
MYSPACE
Social
FACEBOOK
PLAYSTATION
SPACE INVADERS
Games
Password
Username

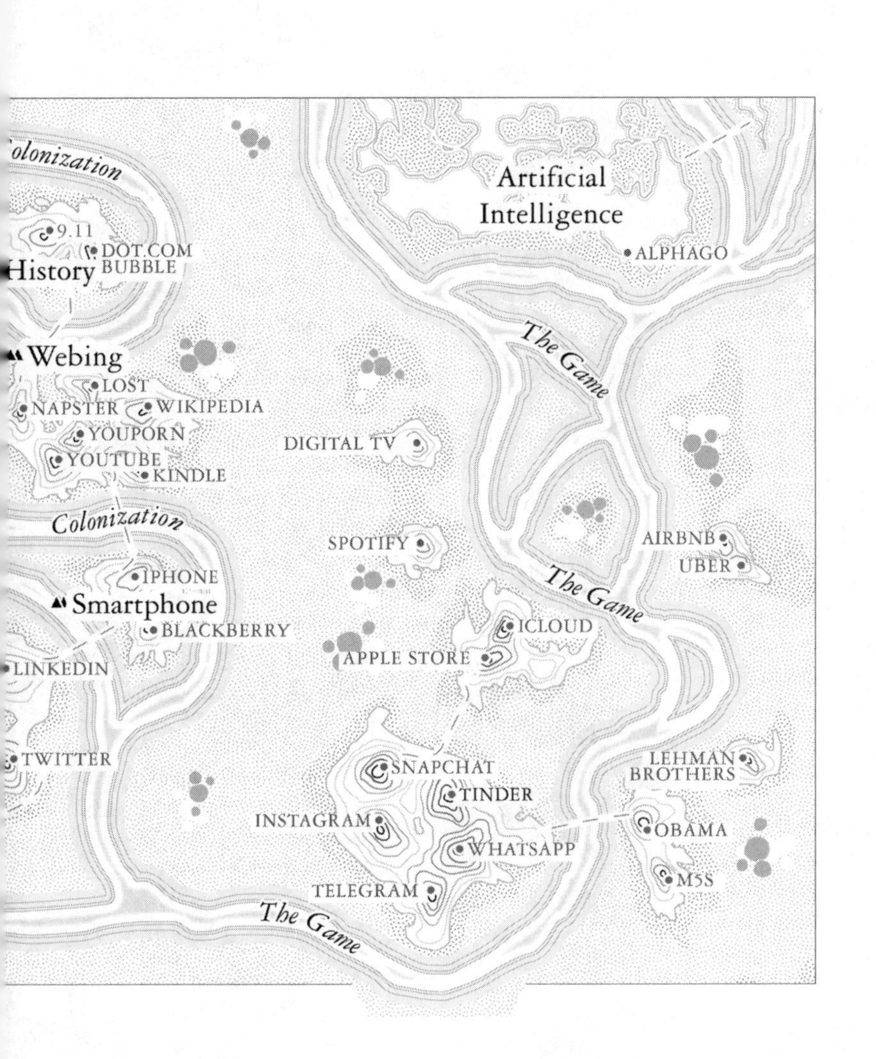

olonization
9.11
DOT.COM BUBBLE
History
Webing
LOST
NAPSTER
WIKIPEDIA
YOUPORN
YOUTUBE
KINDLE
Colonization
IPHONE
Smartphone
BLACKBERRY
LINKEDIN
TWITTER
Artificial Intelligence
ALPHAGO
The Game
DIGITAL TV
SPOTIFY
AIRBNB
UBER
The Game
ICLOUD
APPLE STORE
SNAPCHAT
TINDER
INSTAGRAM
WHATSAPP
TELEGRAM
LEHMAN BROTHERS
OBAMA
M5S
The Game

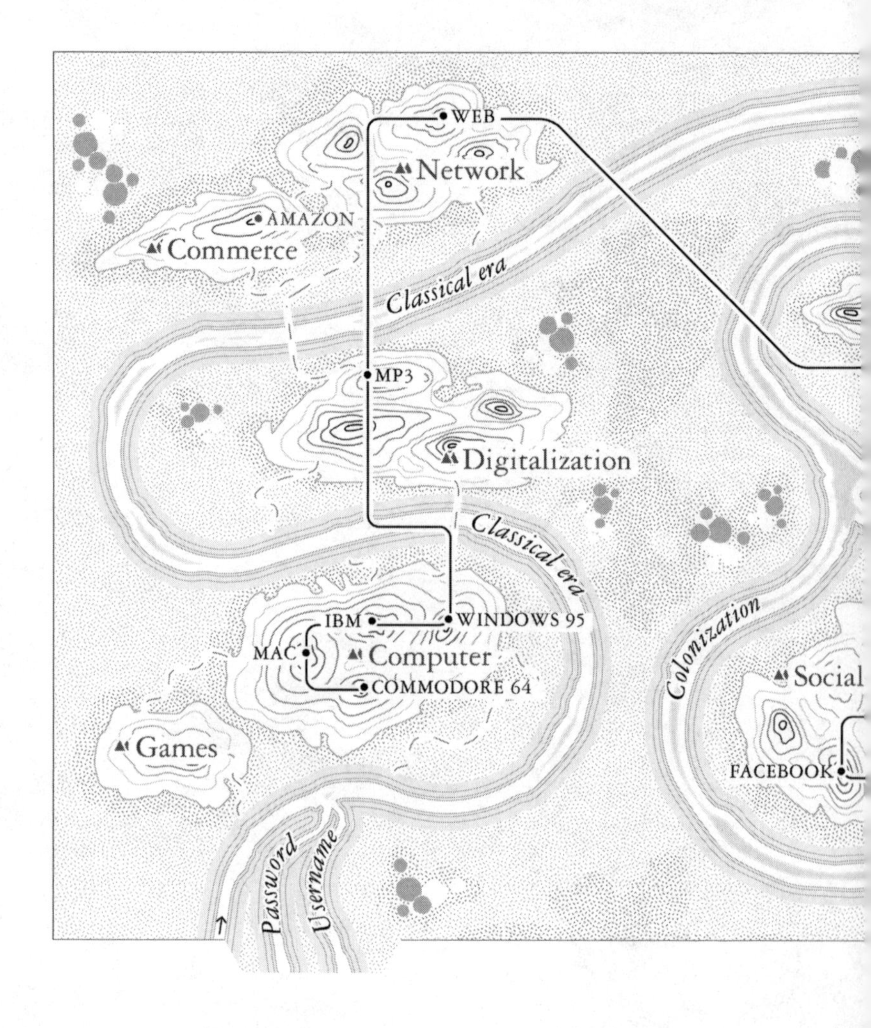
WEB
Network
AMAZON
Commerce
Classical era
MP3
Digitalization
Classical era
IBM
WINDOWS 95
MAC
Computer
COMMODORE 64
Colonization
Social
FACEBOOK
Games
Password
Username

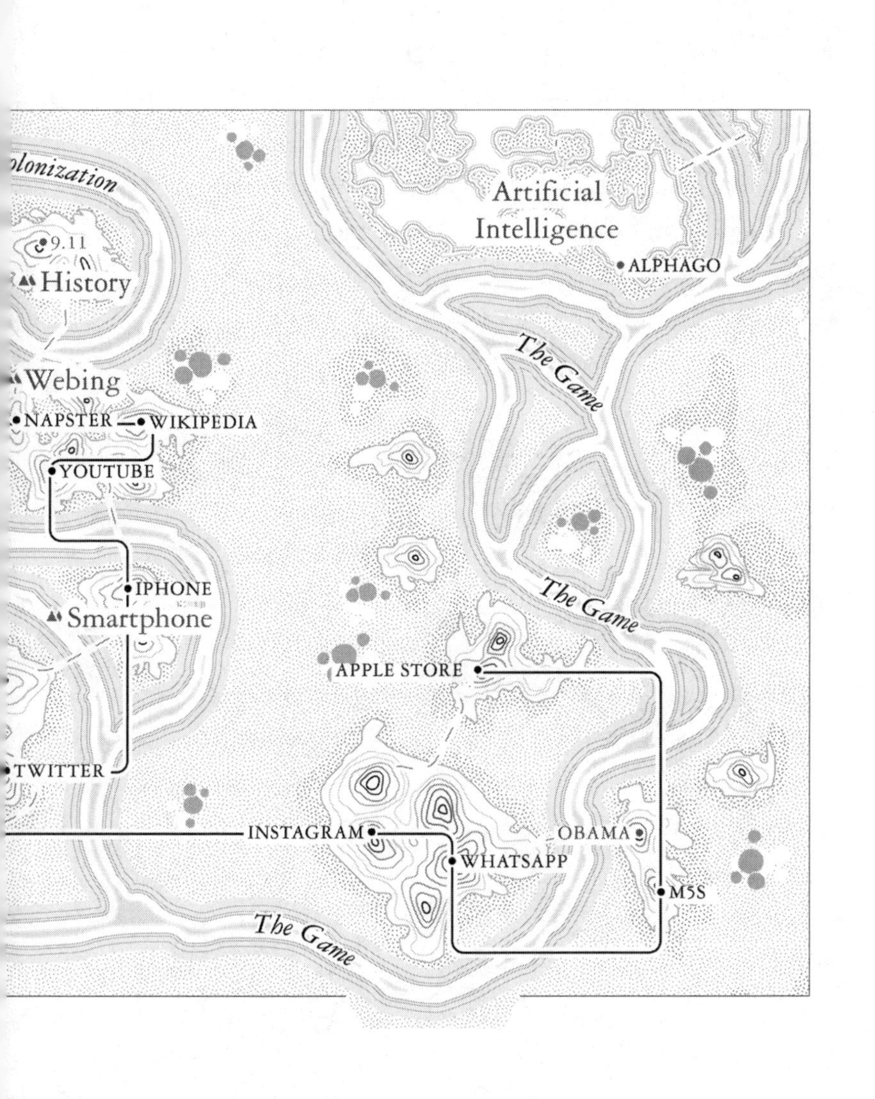

olonization
Artificial
Intelligence
ALPHAGO
9.11
History
The Game
Webing
NAPSTER
WIKIPEDIA
YOUTUBE
IPHONE
The Game
Smartphone
APPLE STORE
TWITTER
INSTAGRAM
OBAMA
WHATSAPP
M5S
The Game

Posture Zero

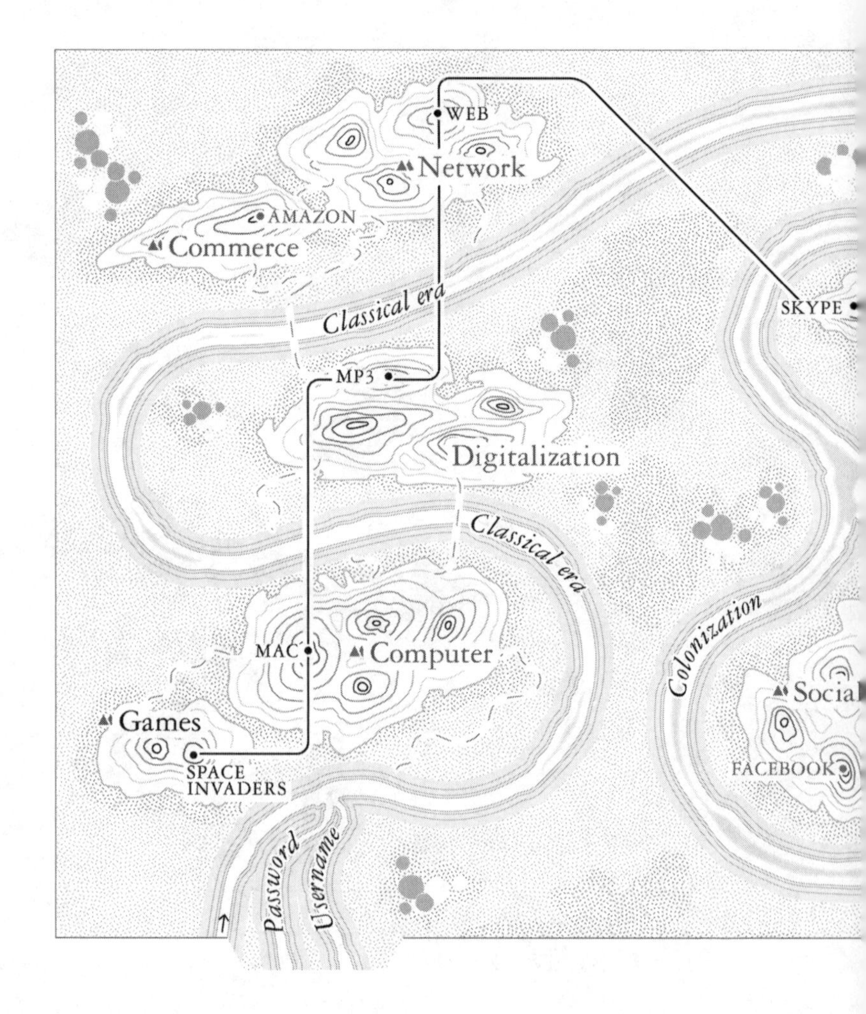

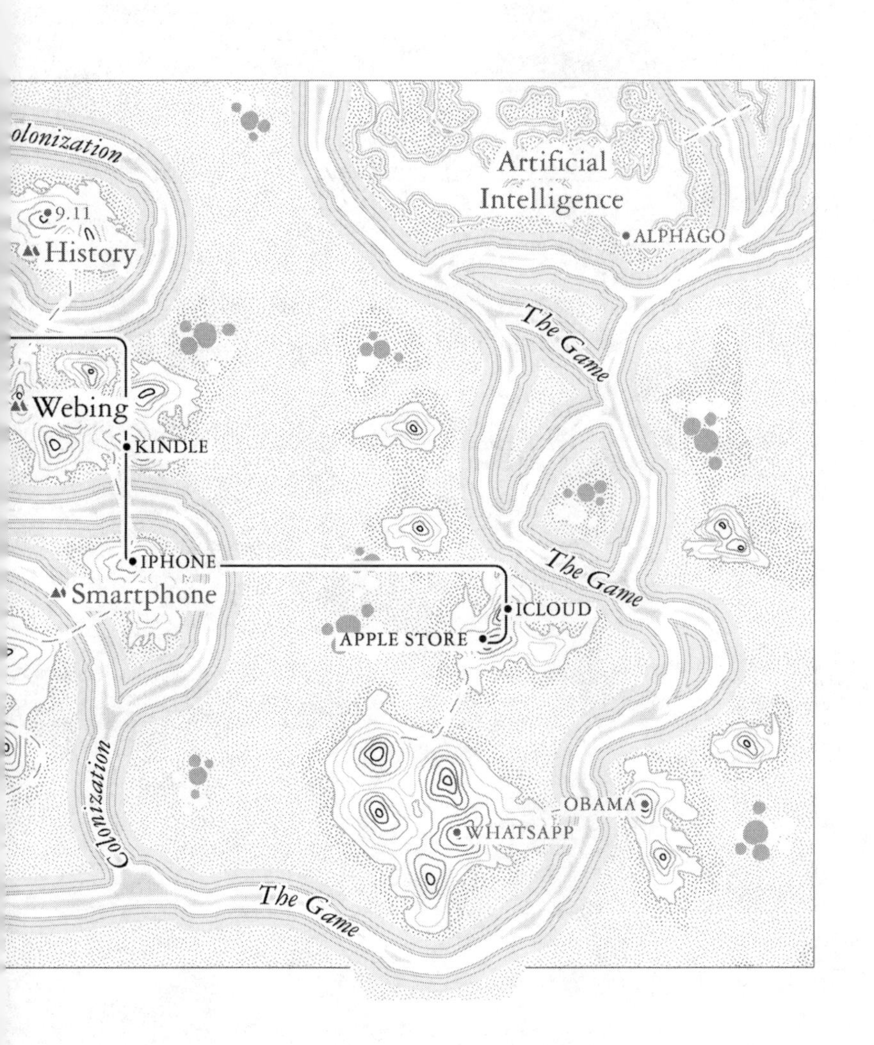

olonization
9.11
History
Webing
KINDLE
IPHONE
Smartphone
Colonization
Artificial Intelligence
ALPHAGO
The Game
The Game
ICLOUD
APPLE STORE
OBAMA
WHATSAPP
The Game

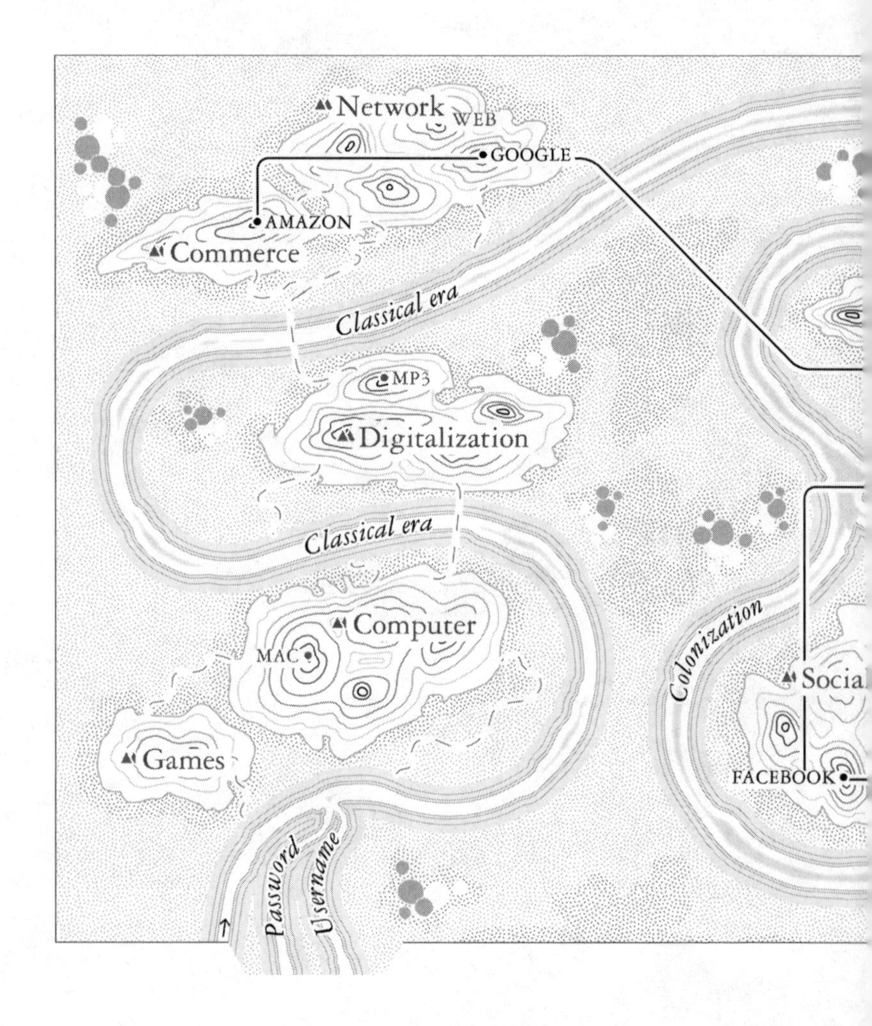
Network
WEB
GOOGLE
AMAZON
Commerce
Classical era
MP3
Digitalization
Classical era
Computer
MAC
Games
Password
Username
Colonization
FACEBOOK

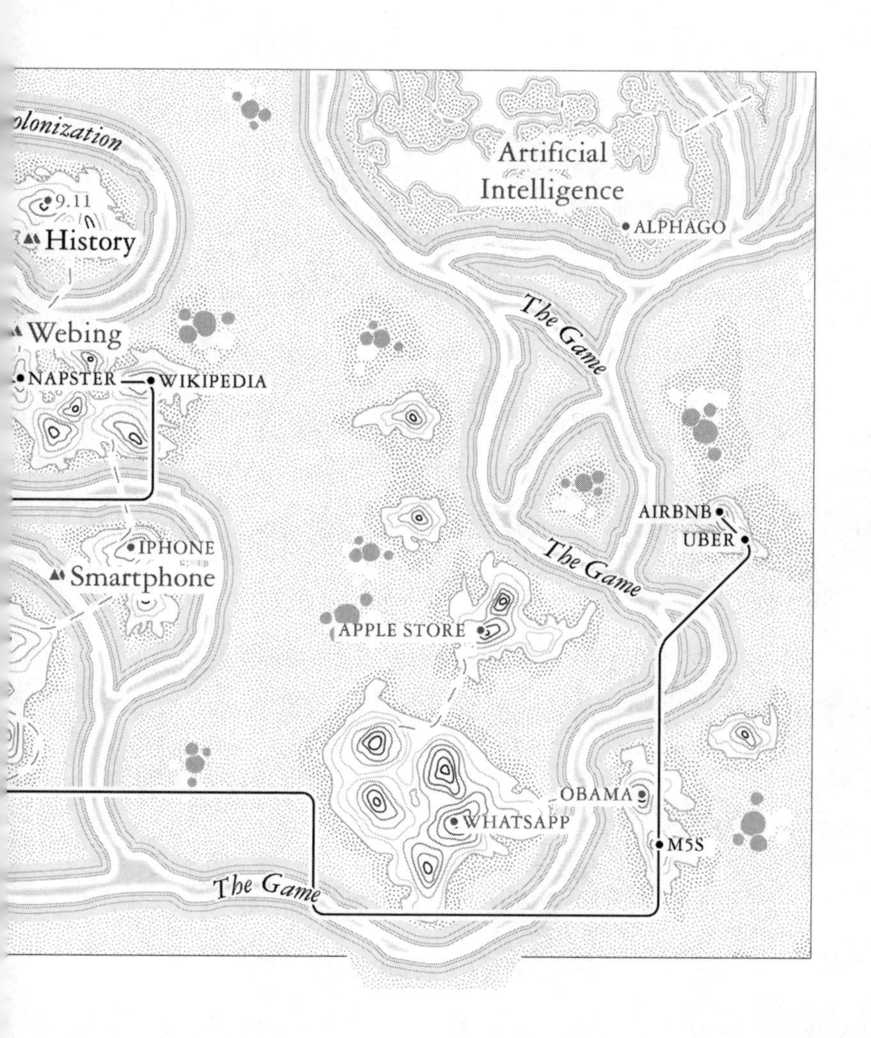

Artificial
Intelligence
ALPHAGO
9.11
History
Webing
NAPSTER
WIKIPEDIA
The Game
AIRBNB
UBER
IPHONE
Smartphone
The Game
APPLE STORE
OBAMA
WHATSAPP
M5S
The Game

Dematerialization

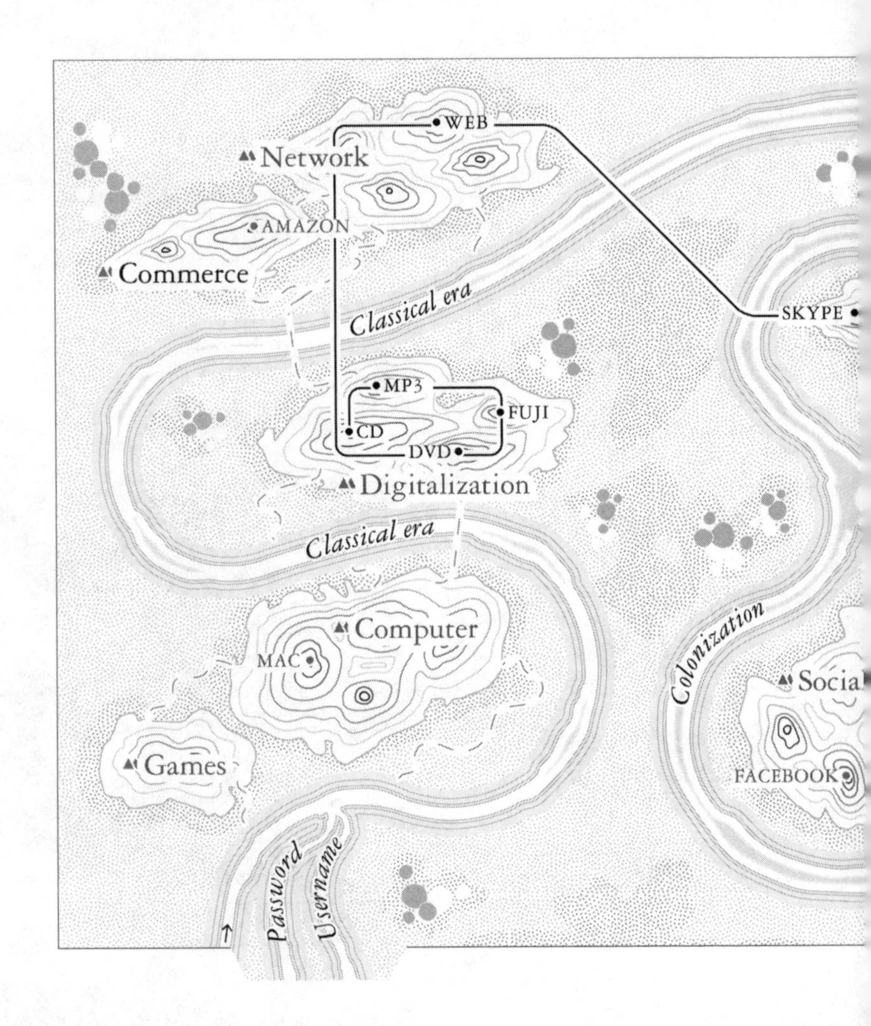

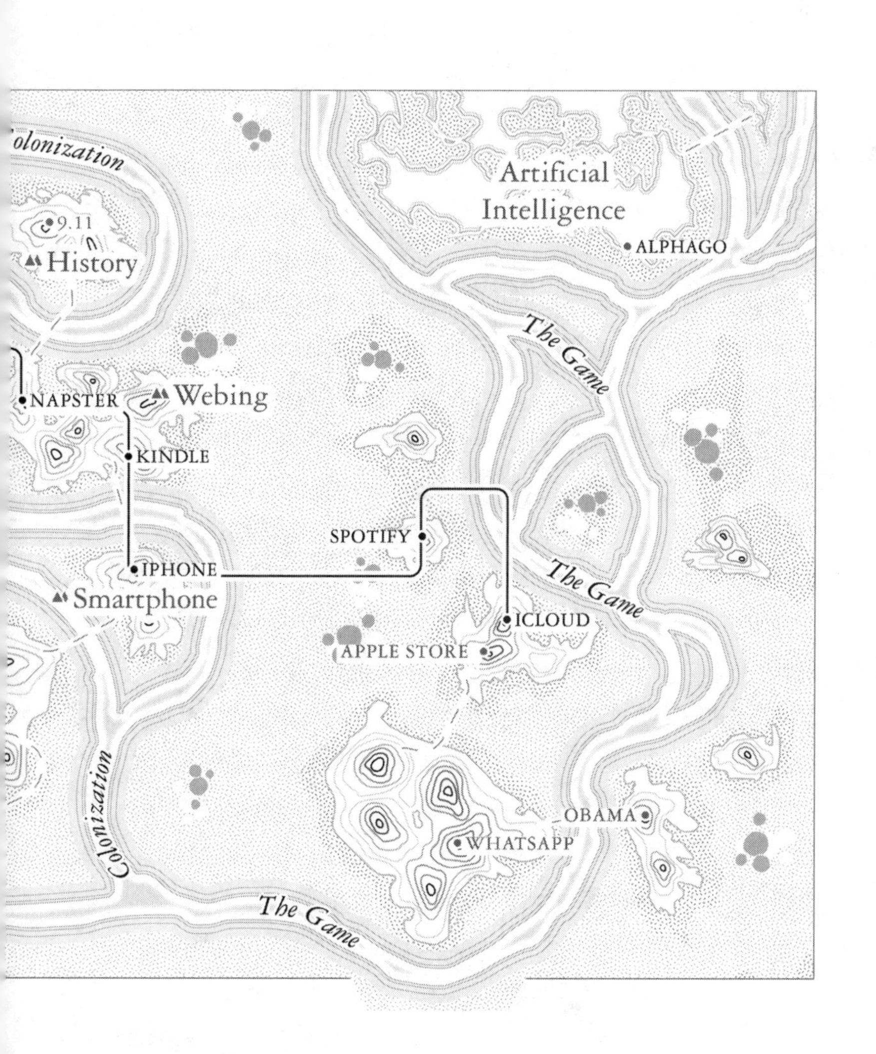

Colonization
9.11
History
Artificial
Intelligence
ALPHAGO
The Game
NAPSTER
Webing
KINDLE
SPOTIFY
IPHONE
Smartphone
The Game
ICLOUD
APPLE STORE
Colonization
OBAMA
WHATSAPP
The Game

Post-Experience

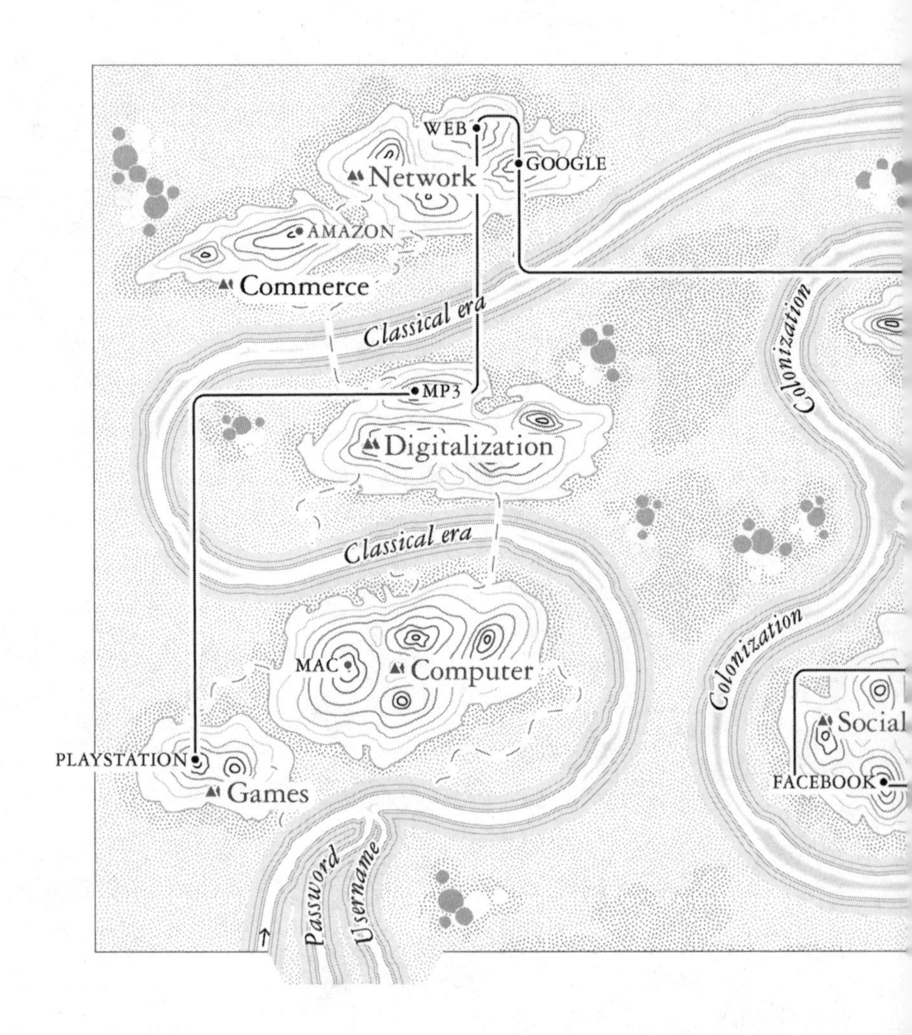

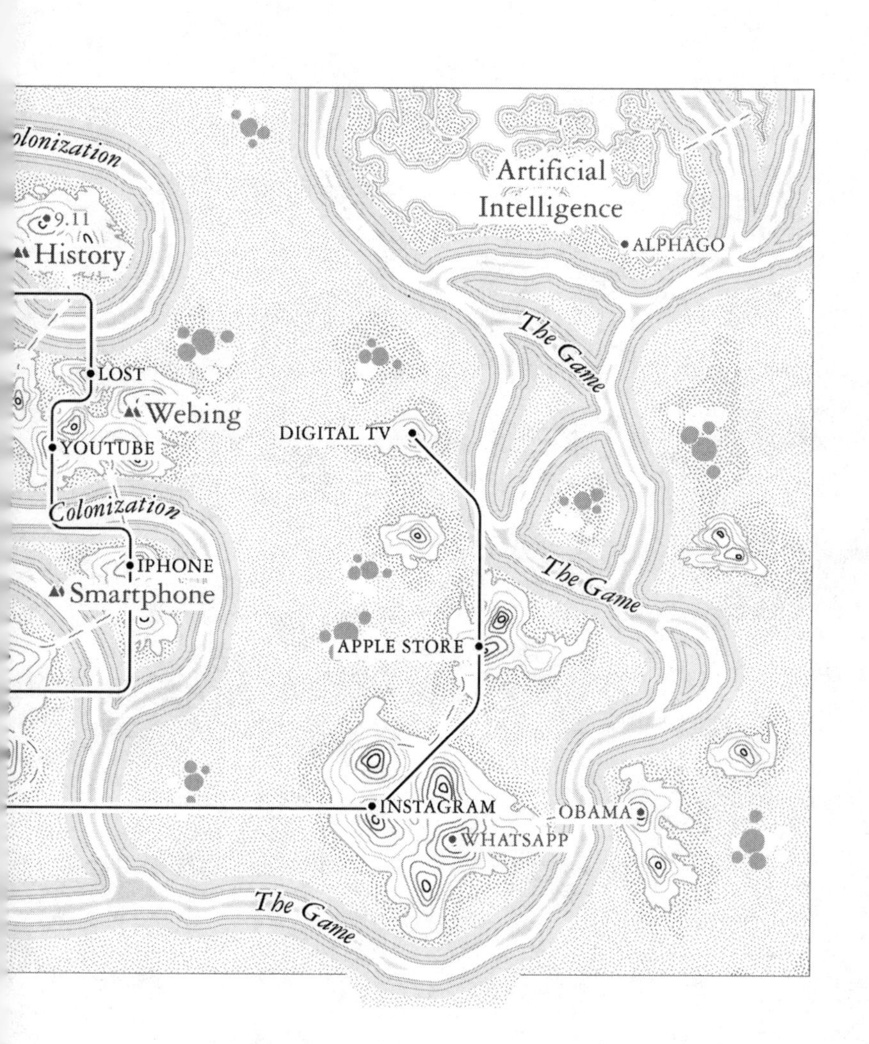
olonization
Artificial Intelligence
9.11
History
ALPHAGO
The Game
LOST
Webing
DIGITAL TV
YOUTUBE
Colonization
IPHONE
Smartphone
The Game
APPLE STORE
INSTAGRAM
OBAMA
WHATSAPP
The Game

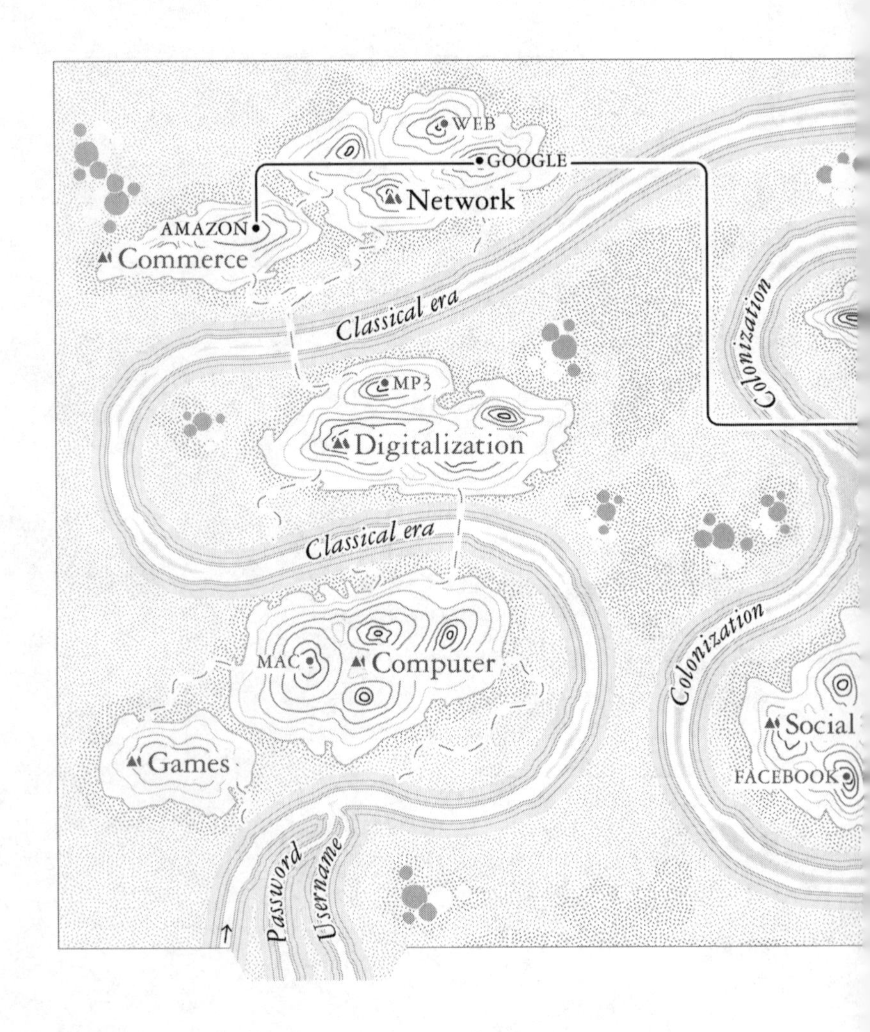
WEB
GOOGLE
Network
AMAZON
Commerce
Classical era
Colonization
MP3
Digitalization
Classical era
Colonization
MAC
Computer
Social
FACEBOOK
Games
Password
Username

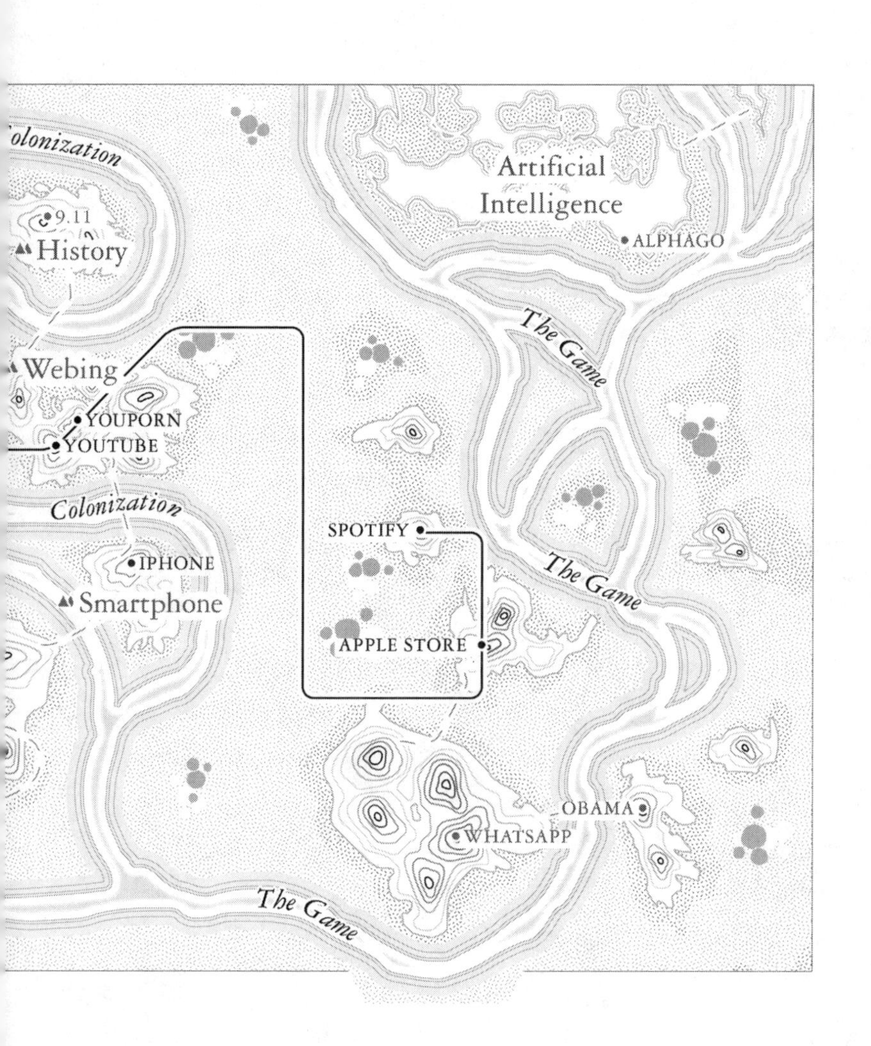
olonization
9.11
History
Webing
YOUPORN
YOUTUBE
Colonization
IPHONE
Smartphone
Artificial Intelligence
ALPHAGO
The Game
The Game
SPOTIFY
APPLE STORE
OBAMA
WHATSAPP
The Game

Username
Password

Play

Maps

▸ Level Up

1. Comets
What remains of the truth

2. Other Otherworlds
What remains of art

3. Contemporary Humanities
What remains to be done

COMETS

What remains of the truth

In the open field of the Game, many things seem out of reach. One of these is the truth.

"The truth" is perhaps overstating it: let's say a verifiable version of facts, a reliable description of events. It would already be quite something to be able to count on a lesser form of truth of this kind.

It is not easy, however. In the Game, something seems to make the truth even more elusive than in the past. After all, if you chose to play on a board where the prime rule of the game was to keep moving, it was never going to be easy to pin facts down with a solid definition. If you have accepted opening up the board to a huge number of players, the daily portrait of the world can only be a composite picture made up of so many different viewpoints that the image is bound to be blurred. If you whiz around the world as post-experience requires, you will soon see that the truth is a sequence of frames where each still, taken on its own, is neither true nor false.

I'll try to put it even more simply. The Game is too unstable, dynamic, and open to be a suitable environment for a creature that is as slow, sedentary, and solemn as the truth.

An example might be useful here. I've selected one that makes me laugh and won't intrude on any of the important issues. It is a little anecdote from a few years ago, at the height of the Era of the Game.

In early 2014, a French gossip magazine published the scoop, complete with glossy photos, that President François Hollande had a young, pretty lover. His official partner at the time was a journalist named Valerie Trierweiler, who took the news very badly and broke off her relationship with the president. She then wrote a revenge book: a ferocious, mercilessly detailed account of her daily life with François Hollande. Publication was announced on September 4 that same year, and public interest was sky high after titillating press stories had kept French citizens on tenterhooks for months. In advance of publication, a few excerpts came out in newspapers, confirming that hell hath no fury like a woman scorned. The title of the book, *Merci pour ce moment* (*Thanks for This Moment*), was sarcastic; everyone knew it would be junk.

Finally, the big day arrived. On September 4, however, in an independent bookstore in Lorient (Brittany), a notice was tacked up in the store window which read: "Trierweiler's book is not available here," followed by a smiley face emoticon. The image of the notice soon went viral on social media and, in no time at all, the store windows of independent bookstores across the country boasted notices announcing things like "Trierweiler's book is not in stock, but you can find books by Balzac, Maupassant, or Proust… " or "We are booksellers. We have eleven thousand books. We don't want to be Trierweiler and Hollande's trash basket." Believe it or not, in a few hours, a mass shift in opinion took place around that first refusal to sell the book. The sarcastic slogan of the growing movement picked up on the title of the book: *No Thanks for This Moment*. Soon, expressions of solidarity began to arrive from outside France. The Game is incredibly fast in this kind of a situation.

In order to appreciate fully the episode, it is important to remember that independent bookstores were fighting a losing battle against Amazon, big distributors, and megastores at the time. Pushed into a corner—somehow victims of the Game—they were folding, one after another. As they were forced out of the market, the sensation was that a certain idea of what bookstores should be, of what book culture should be, of civilization itself was dying with them. This background explains how a rearguard protest against a trashy gossip book became such a powerful symbol in so little time. To put it bluntly, they were mightily pissed off, and as often happens, a minor incident was enough to trigger a revolution.

During this brouhaha, the local Bretagne newspaper (*Ouest-France*) did what a local newspaper is supposed to do: it sent a reporter to interview the bookseller in Lorient, the man who had kicked up the rumpus. I expect they wanted to make a personality out of Damjan Petrovic, the hero of the independent bookstore movement. The local reporter duly asked him why he had put that notice up in his storefront window. And this is the answer he gave:

"The fact is that Valerie Trierweiler's book hadn't arrived yet. People had been coming into the store all morning asking for it, so at a certain point I got fed up and put the notice up."

I can just picture the reporter's jaw dropping. In a last-ditch attempt to save the mission for which he had been sent to the small town, he asked the man if he would have sold Valerie Trierweiler's book if it had arrived in time.

"Of course, why wouldn't I?" he answered.

As the story came out, and in the days following publication of this interview, the *No Thanks for This Moment* movement continued to spread, growing more vigorous, filling many booksellers with pride in their identity and giving them the strength to continue their battle. For a prolonged moment, they all felt like heroes. They were heroes, in every sense of the word. The fact that the trigger for all this had been a misunderstanding was for most of them just a funny detail.

Since the Game remembers everything, we can look back today and find the precise moment when that funny detail started out on its journey. A photo of the notice was posted on Twitter on September 4, 2014, by someone who just happened to be passing by, accompanied by the short comment: "Un vrai libraire à Lorient" (A true bookseller in Lorient).

What this amusing anecdote teaches us is that the Game is itself a slippery slope that facts slide around on without always taking predictable paths. There is no real need for any intervention by powerful players to deviate or invent reality. Facts can travel of their own accord, propelled by underwater currents or tiny anonymous impulses, and once they set sail, it is hard to foresee what path they will take and almost impossible to modify it. Ultimately, the idea one gets is that the Game is made of a *low-density* material that makes it quick and easy to create truths and set them in motion. In the past, creating truth, or shifting it from one place to another, required muscle strength as well as centuries-old expertise. It was a sport reserved for special players in an exclusive club. In the Game, by contrast, owing to its low density, anyone can shift reality. Fabricating it is child's play.

This is causing trouble, as we all know.

For a while now, we have adopted a term that gives us great satisfaction because it gives us the impression that we are controlling the phenomenon: "post-truth." The clichéd phrase is: *we're now living in an era of post-truth*. Let me translate. We are convinced that the Game has led us into a world where the truth is no longer an essential ingredient in forming opinions or taking decisions. Apparently, we have gone beyond truth; we have overtaken facts. We act on the basis of improvised convictions based on nothing or on news that is clearly fake. The force with which these convictions penetrate is attributed to the fact that they are presented as the kind of simple, elementary, compact concepts that Descartes would have called "clear and distinct ideas." Impeccably clever packaging adds strength to their message, especially—people say—where resentment toward elites, experts, and those club members who once had exclusive rights to building the truth is rife. Paying no attention to the truth is a way to catch the old elites offside. It is probably the tail end of a rebellion with roots that go back a long way.

At this point, a question to ask would be this: is the theory of post-truth valid or useful?

Having read pages and pages about the Game, there is one thing we can say with a degree of certainty: the post-truth theory is too elementary to explain what is going on. The Game is neither simple nor childish. In the Game, there is no division between intelligent players, who respect truthful facts, and losers, who only listen to their gut feelings. The idea that, as a result of the digital revolution, a portion of humanity has floated off into a cloud of ignorance and illiberalism, where they are more easily manipulated, does not explain what has happened to truth, to facts, and to the way we elaborate them. Try making sushi with an ax and you'll have more success. In order to explain myself better, I need to go back in time and cite two stories you probably already know.

The first is political. It is a well-known fact that, on February 5, 2003, at the height of the Era of Colonization of the Game, the then secretary of state, Colin Powell, exhibited evidence to the United Nations that was intended to prove that Saddam Hussein's regime was stockpiling weapons of mass destruction. His performance was both theatrical and convincing as he brandished a vial of anthrax. Six weeks later, on the basis of this evidence, the US invaded Iraq, triggering a war that would go on to have incalculable geopolitical consequences in the Middle East and that would cost a great number of human lives. Unfortunately, we now know for sure that the "evidence" Colin Powell presented that day was shamefacedly false. Two years after his UN performance, Powell admitted that the speech would always be a stain on his political career. He claimed that it had been in good faith and blamed the CIA for the hoax. The CIA took it as a compliment.

The second story is more frivolous. Between 1999 and 2005, the cyclist Lance Armstrong won the Tour de France seven times in a row, a feat no cyclist had ever achieved before. Before this exploit, Armstrong had had cancer. The fact that he had not only battled the disease but also gone on to become the greatest cyclist of all time had the irresistible ring of a fairy tale. The myth of his energy and faith in life helped thousands of human beings get up in the morning, however generous (or mean) fate had been with them. A further ingredient of the story was that Armstrong frequently spoke in public about his battle with cancer. He was turned into a hero who had defeated the disease himself and helped other people defeat their fears of it. Sadly, we now know for sure that Armstrong won his seven rounds of the Tour de France because he was injecting himself with steroids like crazy. The doping had gone on for years with stubborn and cunning determination. He denied every accusation brazenly, with admirable impunity, and carried on his career as a hero. Eventually, in the face of overwhelming evidence, he broke down and confessed everything on *The Oprah Winfrey Show*.

* * *

The interesting thing is that, back in the day when these two enormous distortions took place, WE DIDN'T THINK OF CALLING IT POST-TRUTH. The expression existed, it had already been coined, but people clearly didn't find it a useful way to understand what had happened. The term was there, but we didn't know what to do with it. People called the fabrications perpetrated by Powell and Armstrong "lies"—they were no different from the lies people have told for centuries. The expression "post-truth" was there, hanging out in some hidden fold of collective language where it lay snoozing for a few years until it suddenly exploded from under the earth's crust like a geyser. The two forces that brought it to the surface were the Brexit referendum and the Trump election. In both cases, the part of public opinion that was aligned with the dominant narrative, together with the elites who had both forged that narrative and exploited it to govern, suddenly became aware of the mass of outright lies that had circulated during those two political exercises. At the same time, they became concerned by how difficult it was to focus people's attention on facts, or rather, WHAT THEY CONSIDERED FACTS. They couldn't believe people had voted the way they did, and they were so convinced they were right that they were quick to announce the advent of a world where facts no longer counted and legends had taken their place. They didn't, for one minute, try looking at things the other way around. For example, for Leave supporters, perhaps, the FACTS were that their lives were shit and that the idea of trusting their fates to a distant and inscrutable entity such as the European Union was, in their view, irrational and dictated by gut feelings. Most people never bothered to look at it that way. It was more effective to preach about the arrival of a terrible upheaval that would spell the end of civilization as we know it. "Now that we are living in an era of post-truth."

In sum, when we believed Powell's and Armstrong's lies,

everything was more or less normal; when we started hearing rumors that Obama was born in Kenya and not the US, we were sliding down the slippery slope of scorn for facts and trust in gut instincts.

To put it even more brutally, POST-TRUTH is what elites call lies when it is not them telling them. In other epochs, they called them HERESIES.

We don't need to be quite so brutal, however, so let me put it this way. Clearly, the theory of post-truth is the product of an intellectual elite that is running scared and that knows it can no longer control the daily output of truth. The theory cogently chronicles a detachment of the desire for truth from actual knowledge of facts, but then it attributes this split to the irrational drift encouraged by the Game and gives up trying to understand anything further. There is still a certain twentieth-century, static idea in the air about TRUTH that fails to take into account the fact that THE GAME IS TOO FLUID TO PERMIT TRUTH AND TOO ADVANCED TO BE SATISFIED WITH TRUTH. The fact is that, in no time at all, it PRODUCED ITS OWN MODEL OF TRUTH. A model better suited to its rules. It did so by intervening in a specific area that I can only define thus: it intervened in its *design*. What I would like to say is that the Game MODIFIED THE DESIGN OF TRUTH. It didn't disperse it; it didn't change its function; it didn't shift it from the center of the world where it used to be. What it did do was give it a new design. I don't mean "design" in an aesthetic sense; I mean it in its most noble sense. The Game has tinkered with the internal, logical, functional design of truth. It did with truth what Jobs did with phones. There you have it.

In order to try to convince you, I need to go back and examine an object I was convinced would have disappeared forever but hasn't.

THE STRANGE, INSTRUCTIVE CASE OF VINYL SALES

Vinyl records were records made of PVC which, for years (from after World War II to the 1970s), were the most common way to listen to music at home. There were two formats: 331/3 or 45 revolutions per minute (rpm). In the 1970s, sales started to slow down after the arrival of a small object that seemed revolutionary at the time: music cassettes. Yes, the name was pathetic, and the cassettes were not that great either, but they were cheap, you could fit them in your pocket, and you could make your own mixtape with your favorite songs—a bit like making a playlist today on Spotify or iTunes. (I'll open a parenthesis here: teachers in schools should try an experiment. On one side of the class, dub a cassette with their favorite songs; on the other, create a Spotify playlist. Anyone who still complains about the digital revolution at the end of the experiment gets the dunce's hat.) Right, back to what I was saying. In the late eighties, CDs came onto the market, and everybody agreed they were great: they were digital, precise, quick, and attractive. There was one problem, however: they were too expensive. People stopped buying them as soon as cheaper technology came along. Which it did, with a vengeance. Once the MP3 format was invented, music could be stored in a digital form in *compressed files* that were even more immaterial, volatile, and invisible than CDs. Minimum weight, maximum speed. The last time something as magical as this happened was back when elves still existed. Since they occupy no space at all and can be conjured up in no time at all on any of our devices, this format is our chosen method for listening to music nowadays. There is one defect, which is that the quality of the sound is not as good as when it is stored analogically. Not that anybody cares. We are living in a world where we are happy to give up a little quality or poetry in exchange for speed. We were all brought up with a pressure cooker in the house.

Where was I? Ah, yes. Vinyl. With the arrival of MP3 technology, vinyl records were clearly doomed. They stopped producing them. There were a few artisan producers here and there who clung to their art, a little like those craftsmen who make hand-sewn shoes. But vinyl was dead. Amen.

Then there was this amazing piece of news: TURNOVER FOR VINYL IN 2016 HAS OVERTAKEN DIGITAL MUSIC SALES.

Boom!

This news headline is real, I promise. You probably remember it yourself; people were talking about it in bars and at dinner, I swear.

You must understand that when a headline like this comes along, someone like me switches their phone off, drops the kids off at the neighbor's, takes a beer out of the fridge, and starts looking further into the matter. For me, it has the same effect as the pilot of a new season of your favorite TV series (I don't know who is worse off, frankly).

I started looking into things, unbundling the news item, and here are some of the things I found in the article.

As far as we know, it was only for one week (right before Christmas), and only in the UK, that vinyl sales were higher than digital sales. In the US the year before, something similar had happened, but it was not comparable: vinyl sales had been higher than the number of free downloads on sites that earn money through advertising only (YouTube and Spotify's free tier). If we calculate paid downloads (it's cheap, but you still have to pay something), however, the story is different. Vinyl sales were a tenth of digital sales. A more general picture may be useful at this point. If we study the facts, stick to the US market in 2016, and take into account the total sum spent on recorded music, vinyl's market share was 6 percent, while that of digital downloads was in excess of 60 percent. There had been no overtaking.

This is only from the point of view of profits, however. Given that a click to listen to a song on Spotify costs next to nothing and a vinyl record costs anything from $12 to $40, if we count the number of listening hours—the effective presence of vinyl in people's lives—the phenomenon would be even more negligible. Add to this an amusing little statistic provided by the BBC (reporters who get up in the morning to research money flows, bless their souls): half of the people who buy vinyl records and take them home never actually listen to them; a month later, the statistic stays the same. Sweet. It seems 7 percent of them don't even own a record player.

Having said this, the comeback of vinyl is real and surprising. For ten years, sales in the revitalized industry have increased; worldwide, sales of forty million vinyl records are projected for this year. The figures are startling, especially considering that these records are expensive, heavy, hard work to put on, and easy to ruin or dirty; they occupy space, and every thirty minutes or so, they need to be turned around. Of course, these numbers need to be unbundled, too: forty million vinyl records were sold back in 1991, which was the year everyone decided vinyl was dead and that it was crazy to go on producing wax records. When vinyl records were *really* selling well (let's take 1981, the year before the World Cup), they sold more than a BILLION.

Forty million. One billion.

There you have it.

So, let's go back to the headline we started with: TURNOVER FOR VINYL IN 2016 HAS OVERTAKEN DIGITAL MUSIC SALES. Don't make the mistake of sneering in a superior fashion, calling it fake news, and pleading that this is an era of post-truth. Luckily, it's not that simple. It's actually what we might call QUICK-TRUTH.

It's a little communication machine that is highly sophisticated, very popular, and extremely effective. A brilliant creation of the Game. May I explain how it works?

THE BRILLIANT LITTLE QUICK-TRUTH MACHINE

Quick-truth is truth that has been redesigned to be aerodynamic in order to reach the surface of the world; that is, in order to be easily understood and catch people's attention. What it loses in precision and exactitude, it gains in brevity and speed. In fact, it continues losing precision and exactitude until it decides it has gained sufficient brevity and speed to break out onto the surface of the world. When this result has been achieved, it stops; it would never shed a drop of precision more than it needed to. It could be imagined as an animal competing with a host of others to survive, to be known, to reach the surface of the world. What survives is not necessarily the most precise or exact truth; it is the one that travels more quickly, beating the others to the surface.

Take the example of vinyl. TURNOVER FOR VINYL IN 2016 HAS OVERTAKEN DIGITAL MUSIC SALES. Take this headline as the final result of a long journey and follow it back to where it first set out on the trip. If you take the trouble to do this, you will find that one fact is true: millions of vinyl records have been sold in recent years. The news is counterintuitive and teaches us a useful lesson. For a period of time, the information couldn't find its way onto the surface of the world, which meant that nobody was in a position to perceive the trend (vinyl sales had been going up for a decade but nobody was talking about it). Then it suddenly found an opening: one week in the UK when vinyl had a higher turnover than digital downloads. The little animal pushed its way up through the gap. The acceleration was provided by the fact that the truth behind the news had found a perfectly aerodynamic container to push its way through the opening. That is, it was framed in terms of a duel: vinyl against downloads, analogical against digital, the old world against the new world. Duels always attract attention. They simplify things and make them easy to understand. Any news that can be framed as a duel has an easy time in the daily struggle for life. If you think about it, "Achilles against Hector" has been going on for thousands of years. Perfect.

However, the binary frame is not enough. What are the odds that a piece of news about vinyl beating digital downloads in the UK for a week will survive any time at all? Extremely low. In order to be memorable, a duel needs not only the right protagonists (two heroes), it also needs to happen in the right place (Main Street) and at prime time, when everyone can witness it. A little restyling is therefore necessary. Some elements need to be discarded along with a bit of precision. Drop, for example, "for one week" and "in the UK." Do it. Don't argue. We need to overlook the generic expression "digital music." Done. Great.

TURNOVER FOR VINYL IN 2016 HAS OVERTAKEN DIGITAL MUSIC SALES.

The news has come up to the surface. The headline is on the page. Mission accomplished.

Agonizing over whether the news is true or false is not stupid but neither is it urgent or conclusive. In any case, this piece of news contained a truthful fact, and it was its very imprecision that helped some important information rise to the surface of the world: it reported that there was a surprising countermovement, a blip in the straight line of our march toward the future, an unanticipated comeback. This isn't insignificant news. The fact that it was reported enhanced our interpretation of the world, without a doubt. Does it really matter that the information was imprecise? I don't have a sure answer but while I am looking for one, I'm beginning to realize that this piece of (imprecise) news has not only unburied a truth that was worth reporting, but also released other, smaller pieces of news that would never have reached my attention and that have become visible and meaningful only now, in the light of the quick-truth item. I have discovered, for example, that it is not only vinyl records that have enjoyed a comeback; fountain pens too, typewriters and—more importantly—books (soon slippers and carbon paper will be back in fashion) are also selling well. In its belly, the quick-truth headline contained all these truths, and ultimately made them visible by dragging them up onto the surface of the world and shining a spotlight on them so that our attention was grabbed. I have also realized, with the help of this news, that this group of products, a constellation if you like, have in common one easily recognizable feature that I would call: "selling obsolete but vaguely poetic technologies." The rise of this trend makes more people enter the orbit of curiosity about this particular segment of the market (which they had probably forgotten about) and brings them closer to thinking about buying something. This, in its turn, will no doubt revitalize interest on the part of the firms that produce these objects, who will go on to increase their production, which will increase supply and stimulate demand. Money. Jobs. Facts. What was not *really* true in the headline has the chance to become true in the future.

It's amazing how an imprecision can lead to so much meaning and reality. It really can.

If you are shaking your head and wondering what we are coming to or, worse, if you think our new civilization is to blame for the perversion of creating reality out of imprecise truths, then it is my duty to remind you that quick-truth was not an invention of either the digital era or modernity. It is a contraption that was built long ago and has been maneuvered with great skill. Take

Achilles, for example. The one from *The Iliad.* He was considered a half-god: his father was human and his mother was a goddess.

Quick-truth.

It's hard to say today whether the ancient Greeks in the eighth century AD really believed that Achilles was born from the love-making of a man and a goddess. It is reasonable to guess, however, that they didn't have too big a problem with the issue, because the imprecise expression "half-god" handed down something that was absolutely true for them. That is, Achilles was remembered as a man with a strength, violence, madness, and invulnerability they were unable to explain. He appeared to be above the destiny of humankind; they saw in him the disquieting mystery of a potentially invincible nonhuman.

You may object that Achilles was the subject of myths, legends, and poetry. However, this is not the case. At the time, myths and legends were the container information came in. The media were Homer's poems. *The Iliad* was a form of encyclopedia that contained everything the Greeks knew. It was their way of handing down truth. In any case, the formula *half-god* can be found in the history books that replaced myths and legends. From Alexander the Great onward, any aspiring world leader had to introduce himself as a descendent, possibly the son, of a god. Julius Caesar was not the protagonist of a TV series nor the vision of a poet, and yet he was said to have descended from Venus, and he was very particular about pointing this feature out. Nobody would have questioned it. Were they all off their heads? No, they used quick-truth to interpret the world.

We have developed expertise in quick-truth over millennia and yet, if you asked me why it seems so well suited to this era in particular, as if this era had actually generated it, you would be asking an interesting question that you already know the answer to if you have read this book. The answer is that the Game is an ideal habitat for quick-truth. After lying dormant for thousands of years, the concept was given new life there. The idea has always existed, but in the past, it had to find an outlet in high-density systems where news circulated slowly under the supervision of a clutch of experts. News did move, but in slow-motion. The Game has provided the perfect playing field: low density, an unlimited number of players, minimum attrition, high-speed reactions, an infinite selection of paths. A piece of cake. In fact, quick-truth has occupied the center of the pitch and has taken off in terms of status, potential, and power. While throughout the twentieth century it felt like a treacherous caricature of real truth—thought to be based on fixity, permanence, and definition—in the Game, it has taken its revenge. It has shown that, sailing along its unpredictable course that starts nowhere in particular and never ends, its dragnets bring up a huge haul. ITS DESIGN WAS WELL SUITED TO CATCHING AND PRODUCING WIDE SEGMENTS OF THE WORLD. It is important to appreciate the power of quick-truth. Give me a minute of your time and come back with me to the vinyl story.

We had reached the point where an imprecise truth (vinyl sales greater than digital music downloads) expressed a precise truth (obsolete technologies that bring a little poetry into the world have made a massive, surprising comeback, which is on the rise). Well, this is not a finishing point. Quick-truth has traveled a long way, but it hasn't reached its destination yet. The best part is still to come. That is, the beautiful, intoxicating moment when quick-truth starts being interpreted. That is when it snowballs, gathering speed as it rolls downhill. What happens is that there are two ways an item of quick-truth can be interpreted:

1. Human beings are rebelling against technology and are going back to the past;

2. Human beings are so happily advanced in their technological progress that they can allow themselves the luxury of recovering old artifacts and playing around with them because they no longer represent the enemy. It's as if they have a pet python at home; the snake is rendered harmless.

What happens at this point is that the little quick-truth, which has already traveled so far, splits into two divergent downhill paths, leading it to feature either in a magazine that talks about pottery, mountain walking, or yoga, or in *Wired*. Owing to the low density of the magma in the Game, it will continue to snowball in both ecosystems as it meets and resonates with other quick-truths on its path, ultimately forming a heavy, inert entity that will reveal a verifiable constellation of facts. One path will make it seem like a good idea to become an artisanal, old-style cheese maker; the other will create the kind of entrepreneur who opens contemporary caricatures of old-style dairies where you can only pay for your milk with contactless technology.

Thus, going back to that innocent pre-Christmas week in London where it all started, if you trace the journey of our little item of quick-truth all the way to the high-tech dairy (or gorgonzola cheese maker), it is easy to see how much of the world a model of truth like this one is able to create, inhabit, and define. This will teach us to understand and respect quick-truth. You will soon see that its design has a feature you will certainly recognize: it is a journey, not a point in time; it is a figure that unfolds over time, not a fixed hieroglyph; it is a sequence in which every step is fragile but the overall design is strong. You will find in this kind of design the outlines of thousands of other things that surround you in the course of your daily life. Post-experience has the same design. Surfing the web has the same design. There's the logo of the Game in that design.

You will lean over and look more closely at the sophisticated little quick-truth machine with renewed curiosity and increased respect. You will not fail to be fascinated by the fact that this model of truth starts out life as an imprecise fact, a half-truth. You will be struck by its capacity to convert the initial shedding of truth into a strategic advantage as it sacrifices precision in exchange for lightness, speed, agility, effectiveness, even beauty. Movement, diffusion, existence. You may be shocked and think it's a risky strategy, of course. And yet, while you are thinking this, you will also realize that you have already seen it in all the digital tools you use every day. It's the same story as MP3 files, which sacrificed sound quality for transportability; it's the same story as the transition to digital, which sacrificed precision for inconceivable agility. It's the same story as superficiality replacing depth. It's the form of the Game.

Thus, step by step, you will get to the point where you will have to admit that the little machine you are looking at is highly sophisticated, extremely consistent with your way of being in the world, and fantastically well suited to the ecosystem of the Game. It's risky, for sure. We need to learn more about it, for sure. But it is still worth taking seriously, right? Once you do that, you will appreciate that the usual complaints that everything has gone to pot, that facts don't matter any longer, and that we are now living in an era of post-truth present a rather unsophisticated diagnosis. As far as I can see, these complaints are themselves typical quick-truths. They start out life as imprecisions or brutal simplifications, and then they move around perfectly at ease in the Game, their nets dragging the underwater currents, hauling up bits of the world, and then rolling them until they snowball into an inertia where gut instincts are given a name and translated into acceptable thoughts. Hats off. If you are not convinced, all you need to do is try to construct some even quicker quick-truths.

Actually, if I think about it, that's exactly what I'm doing.

FINALE DEDICATED TO STORYTELLING

A quick-truth wins if it manages to get up to the surface more quickly and more effectively than other news. As we have seen, the firmness of its grip on facts is not what counts. It is, rather, its aerodynamic frame that decides its fate. Therefore, if we really want to know what kind of a world we're living in, we need to analyze this frame carefully. WHAT MAKES TRUTH AERODYNAMIC AND THUS WELL SUITED TO PICK UP SPEED IN THE GAME? It's a fascinating topic.

I don't think I've understood enough to give a lecture on the subject, but there is one thing I have a clear idea about, because I've spent a lot of time thinking about it and I know what I'm talking about. OK, I'll tell you. Whatever features make it aerodynamic, and therefore victorious, one of them is more important than all the others. It has a name, and this is STORYTELLING.

Look who's back! Storytelling is one of those features that the Game has brought back to life. It has existed for thousands of years, and now it is cited at every turn. Why? Because the way the Game is designed has provided it with a perfect playing field.

Let's agree on the term before we go on, shall we? *Storytelling*. There is widespread prejudice against storytelling, which is simply a waste of time. People tend to think that there is such a thing as reality and then, beside it, a technique used to talk about it. This is usually seen as the ability to spread a pile of bullshit around, and to do it well.

This is mistaken.

Storytelling does not package, dress up, or trick reality. It is PART of reality; a part of all the things that are real. Would you like a little formula to help you assimilate the idea? I'll give you one. TAKE THE FACTS OUT OF REALITY, AND WHAT YOU GET IS STORYTELLING.

At times, storytelling is just narrative, but very often it isn't. Check out what you're wearing right now: that is storytelling, even though it is not in the shape of a story. Storytelling is like a set of clothes. Like clothes, storytelling dresses what you are in an aerodynamic fashion, so that you can start moving around, connect to other points of the planet, be more easily interpreted, and appear on the index list of reality. Are you the set of clothes? No, you're not. Are you completely separate from your clothes? No, you're not. They are a part of what you are, of the reality you represent, a piece of your real being.

Do you get it? More or less?

Storytelling is a part of reality; it is not always the telling of a story.

Right. Let's go back to quick-truths. Do you remember the ruckus triggered with breathtaking speed by an apparently trivial item of news about a Breton bookseller posting a notice? That was storytelling. One photo with one phrase: "a true bookseller in Lorient." There's a fact there, but unless it meets a storyteller, the fact is voiceless and motionless. It only sets off on its voyage once something gives it a storytelling frame, which turns it into reality. In this specific case, the storytelling component was particularly aerodynamic, as we found out. It was so effective that it ripped the fact clean away from its origins and transformed it into a reality that went way beyond its original intentions. Sometimes storytelling can be an explosive propellant. The low-density fabric of the Game does all the rest (in the twentieth century, nobody would even have known the Lorient bookseller existed).

What about vinyl, then? Do you remember when the news was suddenly thrust under the spotlight after failing to break the surface for years? When for once there was a design capable of the right storytelling: analogical against digital, post against future? It was like throwing a lit match into a pool of gasoline.

What is the lesson here? We have learned that the trajectory of quick-truth is definitely conditioned by thousands of different factors, such as the behavior of its competitors and how rugged and changeable the terrain is. However, its aerodynamics can be almost entirely ascribed to its storytelling reality. I would even say: STORYTELLING IS THE NAME WE GIVE TO ANY DESIGN CAPABLE OF GIVING FACTS AN AERODYNAMIC FRAME TO SET THEM IN MOTION.

Now you will understand why storytelling is everywhere. If something moves, there is storytelling behind it. This has always been true, but in an ecosystem such as the Game, where immobility spells certain death, you can see why it is imbued with greater value. In the Game, nothing survives without storytelling.

This is only devastating news if you cling to the senseless idea that storytelling is a pack of lies that we spread to gild reality. If, however, you let go of this conviction and take storytelling for what it is—a part of reality—it is interesting. It tells us that there is a knack out there in the world for observing and designing the part of reality that is less visible, more hidden, often immaterial, and almost always ungraspable: its aerodynamic component, the way it slices through the air, cuts through the currents, avoids impact, handles lightning speed. This knack saves your life in the Era of the Game.

It saves it from ideas and facts we like but also from ideas and facts we hate. The design in and of itself is neither good nor bad; it is effective and, at times, beautiful. That's it. What we can say is that in the Game, people who know how to exploit it prevail over the others, whether they are called Obama or Trump. The skilled players use it at such an advanced level that there are times when, from the outside, all you can see is their expertise in the apparent absence of truthful facts or ideas with any standing. However, this appearance is deceptive. There is no such thing as a fact without storytelling. If you are lulled by the idea that there are people in the Game who win because of their storytelling skills in a total vacuum of facts or ideas, please continue to be so. I'm not with you. The question is more subtle than you think.

What definitely happened in the Game is that, owing to its low density, the dynamism of truth became more important than its precision. Basically, an imprecise truth with a design well suited for traveling through the Game is more valuable than a slow-moving but precise truth that is stuck in the mud. This verdict may shock you. On the other hand, if you embrace it clear headedly, it is a sign of an intriguing, quite brilliant playing field. What it says is that if I want my facts and ideas to travel, I need to be prepared to give them an aerodynamic profile. I need to hone this until the facts or ideas can slice through the air of collective sensibilities, until it is the right shape and size to snowball in the Game. After all, if someone has managed to pack tons of complexity onto the screen of your iPhone, into a Google algorithm, or into the structure of the web, surely we are not exempt from achieving the same feat. Is it possible that truth is so acute, complex, brilliant, or sophisticated that it won't allow aerodynamism? Even Descartes, in his day, when he was writing the book that would go on to change the path of human thought (*Discourse on Method*), decided to write it in French (the scholars usually wrote in Latin) to make it short and to start out by describing the adventures of his youth. He was trying to make it aerodynamic long before the Game came along. It was the seventeenth century, for God's sake! We can't be so sophisticated that we can't accept a rule even he accepted, right?

During one of my rapid and useless incursions into political life, I once happened to witness a scene I will now describe. There was an issue that needed addressing with various solutions on the table. The problem was which one to choose.

The politician in question considered them all and then asked: "Which story will we be able to tell most effectively?" Mark these words. He didn't wonder which solution would WORK best. He asked which one was most aerodynamic, with the most powerful storytelling potential in its belly. Which one would snowball best in the Game? In a question of this kind, you can detect a detestable form of cynicism, as if the politician were saying, "Who cares about the country? What counts is doing what brings the most votes." However, with a little patience, you can also see an intuition mixed in with the cynicism: an intuition that we are often lacking, that is extremely sharp, and that is in its way prophetic. The politician, if you like, is saying: "Once I've found the solution I like and that fits my value system, I must be cool and detached and choose not only the solution that in theory gives the best results, but also the one that people will understand, accept, metabolize, embody, and contribute to every day when they leave the house. The right solution is useless if I can't get it to roll in the Game. If imprecision guarantees motion, I'll choose it. I'll sacrifice the knight if it helps me get to the center of the chess board." Even a perfect solution that is impossible to describe is, after all, doomed to failure. Worse, it is doomed to lose against solutions that are far worse but well equipped with an aerodynamic coating. Usually those of your adversaries.

Incidentally, this is the problem today with the Left the world over. Supposing they have solutions to people's problems, they are unable to formulate them aerodynamically. Their proposals are dead in the water. There is not a single principle on the Left regarding Europe, immigration, security, or social justice that is endowed with aerodynamism. Their arrogance is unbelievable. The other side, with the populists taking the lead, is incredibly good at the design. I'm not here to judge whether their solutions are more effective or detrimental. The fact is that they design them so that they can be shot like darts in the Game. It's not just their tweets or facile slogans, either. Their aerodynamic nature was shaped elsewhere and a long time before. For example, when they shed the shell of the twentieth-century party and chose a formation with a lighter structure that was better suited to the Game. Or again, when they understood that in the Game, it was impossible to do politics without a leader who strongly, even dramatically, reflected the fact that complexity had to be done away with. A design of this kind is often called populism, but the definition is confusing. In actual fact, it was born on the iPhone screen, on the Google home page, etc. The complexity was hidden below the surface; above the surface, there was simply an icon to tap. A political leader. It wasn't that different with Obama: he had a starkly clear vision of how this mental framework should work. All the others, including Trump, simply followed suit. The Left, generally speaking, does not appreciate the design. It has few talented leaders, and the ones it has, it tends to destroy. It's hard to find decent aerodynamism when constructing things in such an unsuitable form. Finding a remedy after the fact, seeking a good storyteller, or engaging spin doctors to do the job is pretty pathetic. Ideas should be BORN aerodynamic or they will never develop their potential.

Where was I? (There's this thing I hate about politics: it always distracts you from the things that are really important.) Ah, yes. I had just said that Descartes understood that truth without movement is useless, so why shouldn't we appreciate this when we've been trained on all our digital devices? In fact, we are constantly working on quick-truths in our everyday lives. We have become masters of storytelling. We exploit the low density of the Game rather than rejecting it. We practically all know that it is a dangerous system, that it contains within itself the real possibility of constructing effective quick-truths based on almost nothing or on invented facts. Yet, we are learning to control the phenomenon; we are working hard to invent vaccines and antidotes. We practically all realize that we have chosen a system that is highly unstable and that we have forced ourselves to live with brittle truths which are always in motion, condemned to live on a slippery slope. We often suffer the consequences, but an instinctive part of our brain somehow reminds us that when truth or facts have been too firmly fixed, they have led to disasters (disasters we did our best to escape from) to which we have no intention of returning now. The less well equipped, or the over-refined, have been known to give up. However, the central part of the Game is not stopping. It clocks up days in the light of comets it calls truth. It knows how to do so and succeeds in doing so. It will go on doing so with the stubborn and brilliant determination that some kinds of birds show when they migrate toward better lands.

OTHER OTHERWORLDS

What remains of art

As I was saying a few pages back, at one point I started looking more closely at social media. In order to do so, I spent a great deal of time with two people who were much younger than me and who work with social media. As I said before, after speaking with them, I realized that excessive use of social media platforms was not necessarily a sad symptom of digital dependence. They explained convincingly that being present on social media was a way to elaborate on reality, to extract some quality that reality was ungenerous in conceding, to share it with others, and thus to make a piece of theater out of it. They persuaded me to exert caution when I condemned the mass exodus to Facebook and Twitter as a sign that everyone had gone soft in the head. In their view, on the contrary, there was an interesting instinct behind it. That is, the instinct to use digital tools in order to prolong and extend Creation. With their aid, in fact, life didn't have to stop where it stopped; it could be stretched as far as our ambitions wanted it to go.

I had never thought about it that way before, or at least nobody had ever presented it to me as convincingly as these two people did. It was as if a little door had opened in front of me and, while they were showing me how brilliant GIFs are, I stood there pretending to listen to them, but I had actually gone through the little opening to see where it led.

Well, it took me to a place where there was a treacherous little question hanging in the air: if this is the case, *why do I hate being on social media so much?* I mean, if they are a way to elaborate on reality and seek post-experience, basically a way to be alive, why do I hate using them? Worse, *I ask other people to manage my social media accounts.* It's not that I'm cheating or anything. I'm not pretending to be me. It's all perfectly transparent. And yet, I have gotten to the absurd point where I pay others to transport my personality to the otherworld. Why? I stood there while those two people were talking to me and I wondered: why?

Because I'm a damned snob. Tick the box.

Because I was born in 1958. Tick the box.

Because I believe in privacy. Tick the box.

However, with these three justifications I can only get to about 20 percent of the explanation. Trust me. I discovered the real reason while these two people were telling me about the success of memes. When it popped into my head, I found it so interesting that I could have walked off immediately to write down what I finally grasped.

Well, these two people were actually talking to me, so I didn't walk off and leave them there. I'd love to be the kind of person who could do that, but I'm not. So, that evening I didn't write down what I had finally grasped. I piled it all up in a little storage cupboard in my brain for some better occasion.

And now I'm pulling it out. Because this is the better occasion it was waiting for.

I'm not on social media because I write books, perform, teach, and give speeches as part of my work. Once I even made a film; I have often written screenplays. An enormous part of my life is spent elaborating on reality and sending it off to sophisticated otherworlds, where the essence of me is dissolved and recomposed into particles that float along the currents of the collective conversation. I have always lived in a system of reality with twin pumps, but I have always used an older, slower, and far more clunky method than the digital. Therefore, I do not post photos on Facebook; I find it hard to tell stories on Instagram; I don't feel any urgent need to express my views on Twitter. The simple reason is that I have done nothing but post, tell stories, and express my views for years. I did this practically every day, in front of the world, shamelessly using otherworld, old-fashioned apps that existed long before the digital revolution: novels, essays, plays, screenplays, lectures, articles. I guess it's been my privilege—and I've been lucky, too, I suppose. The point, however, is not to appreciate how lucky I have been, but to understand that THE DIGITAL OTHERWORLD IS JUST THE LATEST IN A LONG SERIES OF OTHERWORLDS, MANY OF WHICH ARE STILL HEAVILY POPULATED. I knew this before, of course, but I only REALLY understood it when I crept into that little opening and asked myself why I hate using social media. Every form of social media, from Facebook to *Call of Duty,* stems from things we have been doing for centuries: writing books, fabricating stories, painting pictures, sculpting blocks of marble, composing music. What were we seeking when we did these things? We were trying to complete Creation by duplicating nature and translating it into a language we had coined. We were looking for a way to put what we had understood about life into circulation in a kind of ante-litteram webing. We thus managed to open up the game table, pumping reality into a blood system with two hearts: this world and the otherworld. At times—and we were not so wrong after all—we ended up thinking that the most esoteric truth in the world lay in those otherworlds we created. As Proust said in *Time Regained*: "Real life, life finally uncovered and clarified, the only life in consequence lived to the full, is literature." This is just one example among many. For thousands of years we have believed in the mysterious proximity of beauty and truth, of art and the meaning of life. It is one of our most precious illusions.

Let me sum this up. As suggested in a miniscule but meaningful manner by my allergy to social media—the result of an overdose of my presence in the otherworld in which I work—there must be some logical continuity between the otherworld we always called "art" and the otherworld we now describe as "digital." Let's say that they may be the fruit of the same mental movement, the same strategic move: making copies of the world in languages we have coined. Now what we need to understand is what happened in the transition from traditional otherworlds to the digital otherworld. It's an interesting story because, in that transition, many of the more controversial features of the Game can be clearly seen. It's worth looking into.

Let's try.

Just as we did with *Space Invaders*, we need to go back and look at the games we used to play. Let us go back, then, and look at three otherworlds that used to enjoy great success: plays, paintings, and novels. These were copies of the world written in languages created by human beings: in these formats, they were more communicable, more comprehensible, more usable, and perhaps truer. The technology was neither digital nor analogical. It was called ART.

Plays, paintings, novels. Let's take a look at them as if they had become extinct, as if they had vanished together with the civilization that once used them, like pinball machines in cafés. Let's take a look at them from a distance, from the high plateau of the Game.

Technically, they all had a common design. This is how a millennial might see them:

- The screen was the stage, frame, or book page. (*Different every time? Is that practical?*)

- There was no keyboard. (*That's crazy. You mean, I have to just sit there and watch without being able to do anything else until they finish?*)

- The contents were produced by highly skilled people whose life work it was to write or paint. They were the high priests. After all, these kinds of otherworlds shared some elements with religious practice: temples, rites, liturgies, sacred texts, martyrs, saints, exegetes. (*Jeez…*)

- These objects were only opened rarely and always one by one: you went to the theater to see a play; you opened a book to read a novel. They were thus otherworlds that unrolled gently, with superimpositions of experiences that unfolded one at a time, often after great lapses of time. They were also physically in different places. The theater house was in the square, the painting was in the nobleman's house (and later in museums), and books were in your hand. (*How much time do these people have? Don't they have anything better to do?*)

- They were reserved for the few. Scratch that: for the very few. At the end of the twentieth century, they still required a certain level of income, available time, and education—to the extent that they were often considered the mark of identity for some elites. Owning them conferred identity as belonging to the club. (*Well, congratulations for that.*)

- They were otherworlds you could not access without hard work, application, or, in some cases, further study of one kind or other. They were not always born this way, but the latest civilization to adopt them—the Romantic movement and later the twentieth century—had the upside-down pyramid to respect. They tended to believe that anything valuable required a stint of hard work under the earth's crust. The otherworlds they called *art* were not the exception. (*You mean, I need to study? Are you off your head?*)

- That's it.

To sum up: these otherworlds were expensive, reserved for the privileged few, slow movers, hard to open up, inaccessible, unavoidably linked to the talents of high priests, almost never interactive, and they hardly ever communicated with one another. A true millennial would probably sum them up differently: *They clearly didn't work. Or their batteries needed charging.*

In fact, millennials use other otherworlds that are better built. Millennials can go into these other otherworlds easily, whenever they want. They cost little or nothing; they can be modified, or even created, by means of a keyboard or a console; they can all be reached by means of a single device that can be carried around on their person; they can be shared with others, even when these others are a thousand miles away; and they don't require high priests who know how to do things millennials don't know how to do (not including the programmers, of course, who lurk in the shadows and don't disturb anyone anyway). It is clear that, if the logical, mental, and philosophical habitat we are living in is the Game, these are the features of a functioning otherworld. It makes you wonder how the old otherworlds survived.

Let's wonder, then. It's a matter of understanding what happened to the old otherworlds when the Game first began to dominate, devouring their own sectors as well as many others. Did they end up underwater? Did they resist? Did only the strongest resist? Did they adapt to new habitats? Did firefighters rescue them after the deluge?

It's hard to answer these questions, but we can isolate, identify, and circumscribe a few specific phenomena.

1. One thing that happened is that bilingual buffer zones, as it were, began to form where the old otherworlds and the new lived together: e-books, for example, or Netflix producing movies to watch at home, Arturo Benedetti Michelangeli's piano concerts on Spotify, the streaming of theater events, virtual guided tours of museums. These are all border areas. Cohabitation can, at times, lead to a reduction in quality, but there is a great deal to be gained at the same time. Listening to the Vienna Philharmonic live at the Musikverein is not the same as watching it online, but for most of humanity it is a choice between nothing and something really quite exciting. It's not hard to choose.

   Right. Buffer zones, we were saying. Strategically speaking, they might have looked like a risk for the old otherworlds: the danger was that by demilitarizing the border areas on the frontline of the battle against the Game, the old otherworlds left their flanks undefended against a catastrophic invasion. By way of example, e-books could have totally wiped out printed books. And yet, what actually happened—as we have all seen—was that the creation of these buffer zones was instrumental in calming people's fears. Concerts are still held at the Wiener Musikverein. Actually, it's hard to get a seat, and the quality of the music has not deteriorated in any way. On the contrary, the Game has probably spurred it on and provided it with the tools to do even better. Similarly, we continue to write good novels; sublime opera singers still perform at La Scala in Milan;

people stand in line for hours to see a fifteenth-century Virgin Mary; movie theaters have not gone bust.

2. Owing in part to the bilingual buffer zones, a great many inhabitants of the Game have gained access to otherworlds where they would never have been able to set foot in the past. For years, in theory at least, national politics pursued a similar aim. They claimed they wanted to bash down the gates, which had kept the sophisticated otherworlds under lock and key for years, for the benefit of the gatekeeping elites. The results were disappointing, to say the least. In its way, the Game showed far greater promise: it opened up all the gates and widened access to theaters, museums, and bookstores. A significant number of new faces started to circulate in places where they had never been seen before. They often went in without knocking submissively or asking permission. They just walked in, en masse, weighing in with their numbers and spreading their tastes around. Since they generally came from other cultures—or from no culture at all—the super-refined entities, which were the old otherworlds, started to undergo a chemical process of contamination; at times, they were even poisoned. Some started to suffer (chamber music concerts, for instance), while others quickly developed formidable antibodies. It's hard to take stock, but what is certain is that those who lived in the equivalent of nature parks, where everything had been kept perfectly in order and manically protected, were forced to welcome hordes of visitors who had come to enjoy the view. All that beauty had finally become common heritage, but all that trash these new visitors left on the ground was a terrible pity.

3. In the meantime, some of the old otherworlds started to give birth to organisms that were better suited to surviving the

Game. The best example of this is the transition from movies to TV series. It is a generational passage, as TV series are like digital native movies. They are new creatures, genetically more compatible with the Game. First and foremost, you don't have to go out to watch them. Second, you can watch when you want, how you want, mostly on a device that has a thousand different functions (a movie theater only has one). On a mental level, TV series are movement (typical of the Game), while movies are an action (typical of the twentieth century). TV series are never-ending; their center of gravity is at the beginning, not at the end, exactly like post-experience. In addition, TV series clearly have the structure of a video game—the structure we described when we were looking at the iPhone. They are perfectly suited to the Game. So perfect that it is not illogical to fear for the fate of movies. Well, it certainly wouldn't be the first time a child kills its father in order to become an adult.

4. Another feature that makes old otherworlds unsuited to the Game is the fact that they are chained to the almost high priestly figure of the artist: the creative, the author, the genius—all that stuff. The Game, as we know, doesn't have much truck with high priests. It tends to pulverize power and redistribute it indiscriminately. It breeds millions of individualists who are practically invited to become authors. Do you see where we have a little problem? Take writing, for example. This is a field I know perhaps better than others. There was a time I remember well, when blogs began to proliferate and self-publishing skyrocketed. It felt like anyone could produce an e-book. Social media and the web produced writers who were on the brink of considering themselves authors, and the crisis of authority of the elite was shedding doubt on the allure of bookstores, literary critics, and publishers. Looking around, it was easy to think everything was

about to come crashing down. Well, here's the curious thing. We are still here. The playing field has become harder, for sure. And yet, books are still a field where the uniqueness of a few individuals is recognized, cultivated, supported, and admired. It's a world where there is far more traffic than in the past and far greater vitality. Sure, there's a lot of trash on the ground. Mediocrity travels at the speed of light in dedicated lanes that didn't exist before. There's often chaos, but real writers are still here; they live in certain quarters—not necessarily the most disreputable—and they are free to write good books or bad books. It is entirely up to them. Is the same true of the movies, theater, or music? Perhaps other people should respond, but I suspect that things are not really so very different in these other otherworlds. I would summarize it like this: for reasons I'm not privy to, artists have not been done away with. Even though they are the elite that is even more exclusive and snobbish than others, they are considered a common good. Any idiot can insult them from the comfort of their own home in a blog or a post on social media, but overall the Game loves to need them.

5. The other otherworlds have proliferated, and now one has to elbow one's way to the front to be chosen. Let me go back to my work once again. Many years ago, I probably thought that my main competitors were other writers. Then it became movies, and then TV. Nowadays, I don't even think about how many competitors there are out there waiting to push me back into my old-fashioned otherworld: they are everywhere. Even Zuckerberg is a competitor, though he wouldn't admit it, I bet. (And I'm lucky: at least opening a book is quicker than switching on the device Zuckerberg sells his products on. Imagine playwrights. You have to go out of the house to go to the theater! You have to park!) This competition, of course, pushes you to

perform in an extreme fashion. If there is a lot of competition, you are more likely to scream and shout, exaggerate, or sell at basement discount prices. The result is a civilization where the volume is two bars too high. This is one of the most irritating traits of the Game. It's a civilization of deaf people. Or fools. Or maybe it's on steroids? Yet, I don't believe that's true.

So, here we are. There may be plenty of others, but these are the main things we have seen happen in the collision between the old otherworlds and the Game. At least, these are the main things I am sure actually took place. Is there a clear, coherent picture of the situation? Not really. You can see some of the dynamics that are at play, but we would need a great deal of research to be able to predict further developments or understand their nature. As far as I'm concerned, I feel I can only say two things about the situation that I am totally sure about.

1. The old otherworlds have demonstrated incredible resistance, far greater than expected. Although they are, in theory at least, totally unequipped for the Game, they nevertheless live there permanently, right in the center of town, not out in the sticks. One thing we could add is that they have been the subject of passionate battles, and a great deal of public resources have been spent to shore up their defense. The sensation remains, however, that this would not have been enough to save them if the Game had not had its own good reasons for adopting them, rather than demolishing them. The main good reason is that the old otherworlds guarantee the inhabitants of the Game a heritage of memory, just as religious rites preserve the legacy of a lost homeland for persecuted or exiled populations. Even though the Game is now a stable and victorious

city, it was founded by people in exile who had been on the run. Old otherworlds guarantee continuity between today's reality and yesterday's dreams, between today's well-being and the daring feats of the past, between today's intelligence and the knowledge accumulated over time, between today's homeland and the people who had fled. They provide a past for a civilization that has no past of its own. This is all the more valuable when you remember that the triumph of the Game lies on the shaky foundation of what many believe was an original sin: the decision to place people's lives in the hands of machines. For the inhabitants of a civilization of this kind, being able to prove they directly descended from human beings who were completely human was indispensable. Family trees and genealogies have been drawn up with the precise aim of demonstrating this heritage, an essential element of which was the continuing presence of old otherworlds. We will never be lost altogether as long as we still have books. Not just because of the stories they tell. No, that is less important. It is *how they are made* that counts. They don't have hyperlinks. They are slow. They are silent. They are linear, from left to right (or right to left) and top to bottom. They do not give you points. They have a beginning and an end. As long as we can still use them, we will always be humans. That's why the Game gives books to children. To read once they put down their PlayStation controllers.

2. The old otherworlds have survived relatively well, but the same cannot be said for the elites who used to control them. Writers—who are, after all, wild animals that can adapt to any ecosystem—are doing well, but the entire industry that used to rotate around writers, despite their talent and intelligence, has proved itself to be so unsuited to the Game that

it has inevitably slipped into a twilight zone. Critics in all fields—literary, music, theater, film—are the obvious victims, but the phenomenon has also similarly affected both captains of the cultural industry and commanders of the academic bodies, whose task it was to preserve knowledge and memory. These figures were central to twentieth-century life but, in the Game, they have been marginalized. This could, in part, be seen as part of the more general battle fought by the Game against elites, which would explain the relentless hostilities. But I don't believe this is the case. I believe that, while the old otherworlds were absorbed by the Game, their inhabitants were recalcitrant and refused to do so. The result is that, today, most of the precious things we call ART live within the Game with hardly any protections in place. The legacy of knowledge and intelligence that had guarded it for centuries is all too often immobilized at the margins of the system. Its masters speak a language that has not been translated into the current vernacular; they are out of touch with the most elementary habits of their citizens; their moves are too slow for the Game and, therefore, too stationary to be picked up by the world's radar. A kind of proud fatalism hinders their functioning, with the result that a devastating inertia is sucking them down into a well of oblivion. In a while, we'll forget they ever existed. Thus, works of art are alive and well, but their story is often without a voice. The beauty left to us by our forefathers is in great demand, but it is almost impossible to find, because maps have become impossible to read. A singular interpretation of conservation means that our superb collections are kept under lock and key in order to protect the artifacts from wear and tear. Pedantic rules and regulations mean that the high priests can no longer do miracles while hordes of obtuse believers hold their liturgies hostage and no longer emanate mystery.

The Game is there waiting, with its newborn otherworlds that are brilliant but still babies. They could do with the ancient wisdom of the old otherworlds but, unfortunately, the procedure for tapping into them is not written in a language they understand.

How much is this refined form of foolishness going to cost us?

CONTEMPORARY HUMANITIES

What remains to be done

0. I'm setting down here twenty-five propositions about the Game.

1. The digital insurrection was an almost instinctive move, an abrupt mental twist. It was a reaction to the shock of the twentieth century. The intuition was to flee from that disastrous civilization using an escape route that had been discovered in the first computer science labs. The technology—digital technology—existed, and for the most part, it was used to reinforce the system. But it was easy to imagine that, if the direction of its development were diverted, it could become, on the contrary, an instrument of liberation. The idea originated in a fairly circumscribed community living in California in the seventies: an odd collection of computer scientists, hippies, militant politicians, and brilliant nerds all under the umbrella of a common sentiment. They couldn't stand the world as it was. They were hungry, they were foolish. They were the ones who developed the potential of the digital world by systematically and slyly changing its course in the direction of a fight for freedom. Their first moves defined a match, which all the minds in the world that were working on computer science soon began to play. When the first investments arrived—they soon arrived—the real insurrection got going. The twentieth century was beginning to die without even noticing.

2. The digital insurrection had no ideology, no theoretical underpinning, and no aesthetic. Since it was created mainly by technical or scientific minds, it was more like a collection of practical solutions. Instruments. Tools. It didn't have an explicit ideological basis. It had something better—a method. Stewart Brand summarized it best: "Lots of people try to change human nature, but it's a real waste of time. You can't change human nature, but you can change tools, you can change techniques... and that way you can change civilization." Applied with iron rigor and formidable success, that method has become, in fifty years, the only real ideological principle of the Game. Its sole, quasi-religious belief.

3. The Game can only be understood if we take into account the main purpose of its creation: to make sure a tragedy like the twentieth century will never be repeated.

4. The digital insurrection knew instinctively where the pillars of twentieth-century culture stood and began lucidly and indiscriminately to dig the ground away from under their feet. One by one. We are now in a position where we can outline every phase of that operation and admire its surgical precision. Its two immediate targets were immobility and the dominant elite. Following its methodological principle to the letter, it didn't use theoretical battles or power struggles; it worked by constructing tools. When it had found a certain number of solutions to a given problem, it systematically chose not the fairest or the most beautiful or the simplest: it chose the one that insured the widest margins of movement and cut out the most elites. If you do this tens, hundreds, thousands of times, you'll get some results, I can assure you.

5. The second move was very ambitious: it broke up power and distributed it to the people. A computer on every desk. An otherworld made up of web pages where anyone could circulate, create, share, make money, express themselves—for free. It went so far as to imagine that all the knowledge in the world could be gathered in an encyclopedia written collectively by all humans.

6. It didn't attack the seats of power, it didn't care anything about school, it was indifferent to any church. It dug tunnels around the great fortresses of the twentieth century, knowing that sooner or later they'd collapse.

7. They are collapsing.

8. All this was accomplished using a posture that would later catch on as the logo of that battle of liberation: *human-keyboard-screen*. It was both a physical and a mental posture. It represented the fact that humans had accepted a pact with machines, that they trusted them, and that they were willing to use them as vehicles for approaching the world. It even envisaged a future where machines would become organic prostheses, biological extensions of the human body. Only technical or scientific minds raised in hippy soil could follow a path of this kind without fear, hesitation, or nostalgia. Just one poet in the group would have been enough to block everything.

9. In the late nineties, all the pieces were on the chessboard. Someone made the first move.

10. In the decade that followed, the Game was born. Its founding moment, if we feel we need to choose one, was Steve Jobs's

presentation of the iPhone on January 9, 2007, in San Francisco. He didn't expound theories; he showed us a tool. However, the DNA of the digital insurrection, of which the tool was now fully aware, came to the surface in it and found a form. In that telephone—which was no longer a telephone—one could see the logical structure of video games (the primordial soup of the revolution), the *human-keyboard-screen* posture was perfected, the twentieth-century concept of depth perished, the surface of the world was sanctioned as the home of being, and post-experience was lurking just around the corner. When Steve Jobs left the stage, something had been achieved: the chances that the disasters of the twentieth century would be repeated were temporarily reduced to zero.

11. I hope people who feel nostalgic don't get me wrong. The twentieth century was many things, but over and above everything else, it was one of the most horrific hundred years in the history of humankind, perhaps the most horrific. What made it unspeakably devastating was the fact that it wasn't the result of a failure of civilization or even an expression of brutality: it was the algebraic result of a refined, mature, and wealthy civilization. Nations and empires that had all manner of material and cultural resources at their disposal chose to unleash, for futile reasons, two world wars that they were not able to manage or to stop. The extermination of the Jews was a policy pursued with alarming zeal, and shocking invisibility, on a continent that over centuries had built an awe-inspiring culture. A country that had been the cradle of our ideas of freedom and democracy constructed a weapon so lethal that, for the first time in their history, human beings possessed something they could use to destroy themselves. Finding themselves in a position to use it, they did so without hesitation. Meanwhile, on the other side

of the Iron Curtain, the evil fruits of revolutions, by means of which the twentieth century had dared to dream of better worlds, led to immense suffering, unprecedented acts of violence, and terrifying dictatorships. Is it clear now why the twentieth century is not only the century of Proust but also our nightmare?

12. Whatever one thinks of the Game, that thought is wasted if it doesn't start from the premise that the Game is our insurance against the nightmare of the twentieth century. The strategy has worked, and the circumstances that made the threat of a recurring nightmare credible have now been dismantled. We're used to things as they are now, but we must never forget that there was a time when we would have given anything for an outcome of that kind. Today, if we're asked to give someone our email, we get nervous.

13. In its destruction of the twentieth century, the Game has leveled everything that used to exist without making too many fine distinctions. I repeat: with a guerrilla warfare type of strategy, it left the traditional fortresses of power standing, but when things began to crumble, much was lost, including things that were precious, unique, beautiful, and even *fair*. We are reconstructing some of them, like after a bombing. In some cases, we have returned to how they were before. In others, we haven't. We get the best results when we accept the challenge of using the building materials of the Game and its idea of design.

14. The damage has left a mark across the board; in many cases, it has left a feeling of resentment. The first real war of resistance against the Game was fought peacefully in the nineties. The resistance fighters were for the most part inhabitants of the twentieth century who were fiercely determined not to

abandon their homes. Their opposition was swept away by the overwhelming persistence of the Game.

15. The Game doesn't have a written constitution. There are no texts that legitimize it, regulate it, or lay its foundations. Yet there are "texts" where its DNA is preserved. I'll cite at least five which should be handed down through generations and studied in school: *Spacewar!*, one of the first video games in history (1962); the website where, in 1991, Tim Berners-Lee explained what a website was; the original algorithm of Google (1998); the presentation of the iPhone by Steve Jobs in 2007; Mark Zuckerberg's hearings before the US Senate Judiciary and Commerce committees in April 2018.

16. If you found yourself in the vexing situation of having to save only one of these texts from the Great Flood, the video game should be your first choice. Odd as it may seem, *Spacewar!* already contained the entire genetic code of our civilization, which in fact took its name from similar video games. Video games were the earliest manifestations of the meaning of computers, the potential of digital technology, and the advantages of the *human-keyboard-screen* posture. They demonstrated a certain idea of mental architecture as a collection of physical sensations with a precise idea of speed, a consecration of movement, and an emphasis on a points-system. They were a simplified text that the fathers of the digital revolution could read in order to find out what they were doing and how they could proceed.

17. *Spacewar!* means "war in space." Stewart Brand once wrote, "War in space has done a lot for peace on earth." He meant to remind us that the Game is a civilization of peace. Not so many years ago, we would have given anything to live in a

civilization like that. Today, if we're asked to give someone our email, we get nervous.

18. There was a day, it's hard now to say which, when the Game began to creak under the weight of its tools. If I had to pick one, I would say January 9, 2007. Just as Steve Jobs left the stage.

19. The main issue with the Game is a common occurrence whenever revolutionary movements are prolonged over time. Actions that are ideal for breaking through a front or reversing a trend have long-term consequences that are less ideal. Very simply, often what works in a small community is less manageable when the numbers start to go up. Thus, to take an example, the idea of augmented humanity might be attractive, but controlling the almost inevitable drift becomes problematic: augmented humanity, reinvigorated self-perception, mass individualism, mass egotism. You're distracted for a second, you start going downhill, and the damage is done. Going back and trying to interrupt the flow is as futile as building a dike to control a flooding river. *While it's flooding*, I mean. Yet that's what we have to do. The other possibility would be to abandon the Game. It's not like there's a big crowd at the exit, however.

20. At present, there are three main dysfunctions of the Game which are leading many of its inhabitants to see it as an enemy. The first is that the Game is difficult. It may be fun, but it's too difficult. It's open, unstable, multi-faceted, and never-ending. To survive, you have to have good skills, and these skills aren't taught: you learn by playing, as with video games. The problem is that, unlike in video games, you only have one life, and when you fall, you fall. There are no protective nets, no systems for retrieving the ones who have dropped through the cracks. They

simply end up sliding further and further away. "*No child left behind*" is not a sentiment of the Game.

The second dysfunction is that a system devised to redistribute power has, in the end, distributed potential more than anything else and achieved the unexpected result of creating huge concentrations of power. These are situated in different places compared to twentieth-century concentrations, but they are no less impenetrable. Their logic is at least as opaque as that of the European foreign ministries at the start of the twentieth century. Their financial assets increase at a pace unheard of in the twentieth century.

The third dysfunction lies in the decision to leave the great fortresses of the twentieth century intact: state, school, church. A brilliant gesture with unpleasant long-term consequences, however. In a way, it's as if the Game had left the skeleton of the world intact and then went on to develop lethal muscle masses and a contortionist's joints. It was evident that sooner or later cracks would start appearing at different points and at different times. A collection of micro and macro fractures. To be practical: if the skeleton of education is left to a school that is still set on training good citizens in an average democracy of the eighties, then you can't fool yourself that you're bringing qualified players into the Game: they'll break. Thus, you can conjure up all the mobility possible and continue to develop tools that produce speed, but if the blood system of nations continues to construct bottlenecks, logjams, customs houses, tollbooths, roadblocks, and walls, it becomes hard to vent all that dynamism, pressure, and speed. It is hard to cope with all this serious internal bleeding.

I'd like to add that finding solutions to these three problems is currently unachievable. We can introduce all the correctives we want, and we do so every day, but solutions to problems like these can be found only by minds of the same age. Let me be clear: no one who was born before Google will ever solve any of these problems.

21. The Game is a very young system, so young that it is still produced, in the majority of cases, by people who weren't born there. Brin and Page didn't have smartphones in their pockets when they invented Google, and Berners-Lee couldn't relax at his PlayStation while he was inventing the web. At much lower levels, the detailed daily construction of the Game today is largely in the hands of people who telephoned their girlfriends from a phone booth and used the services of a travel agent when they wanted to book a plane ticket. What we know for certain is that the Game will only unleash its potential when it is designed entirely by the minds that it designed. At that point, it will be itself.

22. I'll offer a single example, perhaps the most delicate. In the past forty years, twentieth-century minds, however prophetic and enlightened, haven't even come close to forging a model of economic development, social justice, or distribution of wealth for the Game. In the Game, the wealthy are wealthy in a very traditional way. The poor, too. It is likely that only a generation of digital natives, who can study the lessons of the past using the tools of the present, will be able to design solutions that do not yet exist today. Invent models, articulate practices, spread culture. It's one of the tasks that lies ahead of them. If they fail, the Game will stay imperfect and, essentially, fragile. Sooner or later, social anger will turn it upside down.

23. At present, the best thing we can do to correct the Game is to make it level. If it were an airplane in flight, we would see it

tilt in one direction: one of the wings aiming at the ground, the other pointing toward the sky. That tilt was there from the beginning and can be summarized in this statistical fact: the overwhelming majority of the Game's founding fathers were white, American, male, and either engineers or scientists. Intelligence in this day and age is more variegated, and the type of flight that allowed the Game to come about is not the best suited for helping it in its maturity. It may have taken engineers to break open the twentieth century and explode it from within, but if *other intelligences* fail to enter the Game's production processes soon, it's unlikely that our future habitat will be sustainable. We need women's culture, humanistic knowledge, legacies that are not necessarily American, talents that grew up in defeat, and minds that come from marginalized areas. If the digital natives presiding over the Game as it matures continue to be white American men who are engineers, the world we live in will be reduced to an endless loop with no future.

24. More than anything else, the Game needs the culture that grew out of Humanism. Its citizens need it for a simple reason: they need to continue to feel that they're human. The Game has driven them to lead an artificial life that may well be suited to a scientist or an engineer but is often unnatural for everyone else. In the next hundred years, while artificial intelligence takes us still further from ourselves, there will be no goods more precious than those which will be able to make people feel human. Absurd as it may seem today, the most impellent need will be to secure an identity for the species. At that point, we'll be able to reap what we have sown in these years.

25. It's not so much the Game that should return to Humanism. It's Humanism that should close the gap and catch up with

the Game. Stubbornly returning to the rites, knowledge systems, and elites that we associate with humanist ideals would be an unforgivable waste of time. Instead, we urgently need to develop a contemporary form of Humanism where the traces humans leave behind are translated into a grammar of the present and become part of the processes that generate the Game every day. We're on to the task. There is a whole area of memory, imagination, sensibility, and mental frames where the inhabitants of the Game have begun to collect and preserve their traces. They don't really distinguish between a fifteenth-century philosophical treatise and a mountain path. They are looking for human remains, and wherever they find them, they record them. They discard some of the evidence and preserve many other clues. They translate everything with the clear intention of constructing the Game in a way that is fitting for humans. Not only *produced* by humans: *well suited* to them.

They are levelling the flight of the Game.

The Great Library where they're engaged in this task doesn't exist, and yet it is everywhere. Its catalogue is immense: you could spend a lifetime just scanning the titles between *The Simpsons* and *Spinoza*. Anyone who went there would make magnificent discoveries: a certain capacity to match colors or combine tastes, the possibility of stringing together long thoughts or sentences, a mysterious ability to do things slowly or to stand still. The risk that they remain fossils to be admired in museums on Sundays is there, of course. But if they become *contemporary humanities*, that is, scenes from the Game, we'll find ourselves playing them, and then it will be another story entirely. A story of humans, once again.

ACKNOWLEDGMENTS

I am particularly grateful to Annalisa Ambrosio and Elisa Botticella, who helped me in the research for this book. It's not just that they are cultivated, attentive, and brilliant. What I love about being with them is that we have a really good laugh together.

In different ways, and sometimes without even knowing it, many people helped me write *The Game*: Sebastiano Iannizzotto, Valentina Rivetti, Martino Gozzi, Arianna Montorsi, Riccardo Zecchina, Marta Trucco, Riccardo Luna, Federico Rampini, Gregorio Botta, Valentina Desalvo, Marco Ponti, Dario Voltolini, Tito Faraci, and Sebastiano Baricco. My thanks go to all.

Luigi Farrauto and Andrea Novali were fantastic travel companions.

ABOUT THE AUTHOR

ALESSANDRO BARICCO was born in Turin in 1958. Baricco's novels have won numerous literary awards, including the Prix Médicis étranger in France and the Selezione Campiello, Viareggio, and Palazzo del Bosco prizes in Italy. The film of *Seta* (*Silk*), directed by François Girard and starring Michael Pitt and Keira Knightley, was released in September 2007, and Baricco made his directorial debut with *Lecture 21*, which premiered at the 2008 Locarno film festival.